The Logic of Theoretical Sociology

"While highlighting the shortcomings of inductive and applied social science that has us heading in the wrong direction, Willer and Emanuelson provide a guide for not just acquiring a collection of contextualized results, but general cumulative knowledge. *The Logic of Theoretical Sociology* is an important corrective that explains through example how to develop sociological knowledge."

—David Melamed, *Professor of Sociology & Core Faculty in the Translational Data Analytics Institute, The Ohio State University*

"The Logic of Theoretical Sociology traces the development of social research from its philosophical foundations to its current, decidedly non-philosophical, incarnation. It shows both why social science is so hard to do scientifically and how to do it. Important and timely given the crises of confidence in social and medical research, Willer and Emanuelson's book gives us hope for truly scientific social science because they have done it themselves."

—Michael J. Lovaglia, *Department of Sociology and Criminology, University of Iowa*

"The authors of this book make a strong case against empiricism and for the use of theory in all sciences. Instead of seeing the plurality of theories in sociology as a problem, they view it as a possibility. With an elegant example, they show in chapter 9 how a conjunction of theories may provide a satisfactory explanation of the rise of the state."

—Lars Udehn, *Stockholm University, Sociologiska Institutionen,* 106 91 *Stockholm*

"Willer and Emanuelson, with precision and clarity, provide a roadmap for building and testing theories in sociology. Theory building has a precarious position in sociology today, and this book, with significant and important historical grounding, shows how we can right the ship. It is an important read for anyone concerned with the current state of building knowledge in sociology, as well as those who are just learning the craft of sociological theory building and research."

—Lucas Jeff, *University of Maryland*

"In a time of seemingly unlimited faith in "big data" and in which any kind of thinking, let alone of the critical kind, is increasingly outsourced to AI, this is a very important book. Willer and Emanuelson encourage the reader to take a step back and contemplate the current state of "theoretical sociology". They reestablish the deep and valuable links between sociology and other theory-driven sciences, and reevaluate the experimental method. This is an excellent place to start (and come back to) for any serious student of society who aspires to truly understand the topics they study."

—Jacob Dijkstra, *Director of Studies, Department of Sociology, University of Groningen, The Netherlands*

David Willer • Pamela Emanuelson

The Logic of Theoretical Sociology

An Historical Assessment

palgrave
macmillan

David Willer
University of South Carolina
Columbia, SC, USA

Pamela Emanuelson
North Dakota State University
Fargo, ND, USA

ISBN 978-3-031-91708-0 ISBN 978-3-031-91709-7 (eBook)
https://doi.org/10.1007/978-3-031-91709-7

This Palgrave Macmillan imprint is published by the registered company Springer Nature Switzerland AG.
The registered company address is: Gewerbestrasse 11, 6330 Cham, Switzerland

Forward

I am a sociologist because of David Willer, the first author of this book. The story of how this came to be is one for another medium. The important point here is that Dave's commitment to building a scientific sociology—and an equally strong commitment to training others how to do the clearest work they could do–was central to his training of me as an MA student.

A bit after my PhD, I accepted a position at the University of South Carolina as an assistant professor of sociology. Besides the joy of being Dave's colleague and being embedded in the culture of scientific work that he had built around the experimental laboratory in the department, Pamela Emanuelson, the second author of this book, arrived as a graduate student in Dave's lab around the same time. Pam was—and still is—among the very best graduate student alums in the department. And like Dave, she had an unwavering commitment to sociology as a science. It makes sense that they would eventually write this book together.

The Logic of Theoretical Sociology is a practical guide to theory building in sociology. You might ask: Don't these already exist? What's different about this one? There are lots of things. Here are a few. Combined, the authors of this book have been doing two important things for three quarters of a century: (1) attempting to clarify and promote the value of scientific sociology, and (2) actively building scientific sociological research programs. The fact that the authors have so much experience *doing* scientific sociology makes this book far more effective in its goal of teaching readers how to build, refine and test sociological theory than alternatives out there, many of which are written by philosophers of science with no such direct experience. As someone who teaches theory construction and analysis to graduate students, this is perhaps the key thing that sets this book apart from other readings I assign.

Another key distinction is that this book situates contemporary sociological theory building not only within the sociological classics (especially Marx and Weber) but the physical sciences as well. The authors continually and usefully draw parallels between the work of Aristotle, Galileo and Einstein and classic and contemporary sociology. This has two advantages. First, the authors situate issues of theory building in a much broader context than readers might otherwise get. Early on, for instance, the authors trace a fork in the road where sociology could have taken a more scientific path. Specifically, sociology (whether explicitly or not) mostly took an inductive path. But Karl Popper showed in the 1930s that there is no inductive *principle,* and Willer and Emanuelson further show that there is no inductive *methodology*. Given this, the production of general knowledge requires theory. (Much of the rest of the book is devoted to how sociology can correct course.) Second, this context allows the authors to convincingly dispel the commonly held belief that sociology cannot be scientific because—unlike physics, for example—its subjects think and react. The authors show that the ways sociology is unique from other sciences do not limit its potential to develop as a science.

The book makes several other particularly noteworthy contributions. For one, the authors point to the role of theory in the "replication crisis" i.e., the discovery that many well-known research findings cannot be replicated. As many readers know, psychology has been dealing with a replication crisis for some time and there is no reason to expect that findings from sociology would fare much better. As the authors note, psychologists have (not wrongly) sought to address the crisis on two fronts: (1) attempting to weed out bad actors who, in a few particularly egregious cases, have wholly fabricated their data, and (2) shoddy, but common, research practices (e.g., waiting until the results are known and then creating hypotheses that "predict" the already-known findings). But Willer and Emanuelson argue convincingly that limiting bad actors and tightening research practices will not resolve the replication crisis if social scientists do not begin to rely on theory to design research studies. This book teaches readers how to do this.

Additionally, a common critique of experimental research in sociology is that it is—at best—limited to advancing our understanding of small groups. What about the messy real world with many actors? What about big historical events? The authors show why these critiques are misguided. Experiments, the authors argue, are not intended to be generalized to the real world. Rather they are tests of theories. And it is theories that have been tested, refined and extended that can be then used to explain much more complex phenomena. Indeed, the authors show how one social theory, Elementary Theory, when used alongside other theories as components of a "theory toolbox" help explain major historical transformations like the development of the political state and the rise of feudalism in Europe.

Since the goal of the book is to show readers how to do more scientific research, it is important to note that one walks away from the book feeling optimistic that this can be done. The authors give concrete examples on the development of theory and a theoretical research program as part of this guidance. And while their book promotes an *ideal* of scientific sociology, it recognizes that there is a continuum of scientific research in sociology. Though I thought I knew most of what the authors had to say on the topic, I walked away with many ideas about how to move my own research closer to the ideal advocated and practiced by these authors. I am confident other sociologists will as well.

Columbia, SC, USA — Brent Simpson

Acknowledgments

We offer our heartfelt thanks to those who contributed to the completion of this book. To Pat Willer, thank you for your invaluable assistance in editing. Your insights and careful attention to detail brings clarity to our work. To Lars Udehn, thank you for your thoughtful feedback on the manuscript and for allowing us to bounce ideas off you. Your contributions were instrumental in refining our arguments. Additionally, thank you Brent Simpson for your generous Foreword, which added an extra layer of enthusiasm to our book.

Their support and contributions have made this work stronger, and we are truly appreciative of their time, expertise and encouragement.

Competing Interests The authors have no competing interests to declare that are relevant to the content of this manuscript.

Contents

1 **Introduction** 1
References 9

Part I Empiricism and the Stagnation of Social Theory 11

2 **The Fork in the Path** 13
John Stuart Mill 14
Karl Pearson 17
Ronald Fisher 20
Conclusion: The Path Taken 27
References 29

3 **Social Science and Its Myths** 31
Robert Merton 34
Carl Hempel 35
Karl Popper 38
Thomas Kuhn 41
Conclusion 44
References 45

4 **The Replication Problem** 49
Fisher Design and the Problem of Induction 52
Theory-Designed Experiments 52

Hybrid Experiments and Scientific Exploration 57
Replication Success by Type of Experiment 60
Experimentation in Economics 61
Falsification as a Process 62
Conclusion 67
References 67

Part II Theory-Driven Research in Social Science 73

5 Worldviews 75
The Four Components of Scientific Theory 75
Aristotle 76
Galileo Galilei and Isaac Newton 78
Karl Marx 82
Max Weber 85
More Recent Worldviews 88
Conclusion: Worldviews 95
References 96

6 Contemporary Theories and Schools of Thought 99
Balance Theory: Modeling Sentiments 100
Game Theory 102
Rational Choice Theory 106
Social Network Analysis 108
Power-Dependence Theory 110
References 114

7 Elementary Theory 117
Social Actors 118
Actors in Relationships 122
Bargaining and Resistance 122
Structural Power I: Exclusion 125
Structural Power II: Inclusion 127
Structural Power III: Some Mixed Cases 130
Coercion: Relational and Structural Power 131
Strategic Analysis 133
Conclusion 136
References 137

8 Theory and Experiment 139
One Fisher Design and Four Theory-Driven Designs 141
Elementary Theory-Driven Experiments 150
Conclusion 154
References 156

9 Theory Toolbox 157
Development of the State 160
Theory Toolbox I 171
The Transition to Feudal Social Structures 172
Theory Toolbox II 184
Interdependent Political Structures 184
Theory Toolbox III 195
References 195

10 Conclusions 201
References 209

Index 211

List of Figures

Fig. 2.1	Stouffer's design for an ideal experiment	25
Fig. 4.1	Models of DNA replication	54
Fig. 4.2	An Experiment on a Tiered Exchange Structure with Mobility	56
Fig. 4.3	Models of Virulent S-Strain and non-Virulent R-Strain *Streptococcus pneumoniae*	58
Fig. 4.4	The theory-experiment feedback loop	63
Fig. 4.5	An exchange structure	64
Fig. 5.1	Galileo's diagram of projectile motion	80
Fig. 5.2	Newton's diagram for the effect of Joint Forces N & M	81
Fig. 5.3	Newton's action of centripetal force	82
Fig. 5.4	Coleman's diagram for the effects of religious doctrine on economic organization	92
Fig. 5.5	An overfull labor market	94
Fig. 5.6	Formation of capitalist classes in Coleman's Boat	94
Fig. 6.1	Heider's six diagrams of cognitive balance and imbalance	100
Fig. 6.2	Cartwright & Harary's generalization of Heider's graph	102
Fig. 6.3	One State of the Prisoner's Dilemma	103
Fig. 6.4	Three notable games	105
Fig. 6.5	Coleman's Boat	106
Fig. 6.6	A 6-node Star Network	109
Fig. 6.7	Cook et al. (1983) Figure 1c, A 5-Person Network with Three Positions	111
Fig. 6.8	A Structure to Falsify Vulnerability (Willer, 1986, Figure 2)	112
Fig. 6.9	Three weak power networks	114
Fig. 7.1	Models of sanctions and relations	118
Fig. 7.2	Preferences for outcome by actor and relation type	120
Fig. 7.3	Two forms of exchange relations	122

Fig. 7.4	An exchange and a coercive relations each with transmitting and receiving signs and sanction values	123
Fig. 7.5	Two null exchange structures	124
Fig. 7.6	A 2-branch exclusive exchange structure	125
Fig. 7.7	Sequential exchanges for ordering and inclusion in a 2-branch network	128
Fig. 7.8	Centralized coercive structures	133
Fig. 7.9	A strong power exclusionary exchange structure	134
Fig. 7.10	A defection chain	135
Fig. 8.1	A Br32 exchange network	151
Fig. 8.2	A 7-position strong power networks and its computer-mediated experiment replica	152
Fig. 8.3	A centralized coercive structure	153
Fig. 9.1	Simple, Patrilineal Status-Lineage Structure	162
Fig. 9.2	Indirect Coercion of Commoners by an Aristocrat	164
Fig. 9.3	Direct Coercion of Commoners by an Aristocrat	166
Fig. 9.4	A Status Lineage with Downward Mobility	168
Fig. 9.5	Assembled component power structures to build a model of bureaucracy	182
Fig. 9.6	Competing power and status hierarchies	189
Fig. 9.7	Connected Senior and Cadet Lineages	191

List of Tables

Table 4.1 Frequency of replication by the use of model and replica for designing the experiment 60
Table 8.1 Prisoner Dilemma payoffs for a 2-person game 143

CHAPTER 1

Introduction

The central theme of this book is that research methods of sociology, having failed to produce science, need to be replaced by scientific methods that can and do explain social phenomena. Scientific methods have been shown to produce the kind of knowledge that grows and cumulates. At the core of those methods is scientific theory, for theory alone can explain, predict, and cumulate. None of this is to suggest that good science is not now being practiced in sociology. To the contrary, as we will show, theory is being tested and its scope is growing as seen in papers scattered across the journals. Here we will examine the theoretical methodology underpinning this work. Our hope is to encourage further growth.

By the middle of the twentieth century there was a fork in the path of development of sociology as a science. One path led toward science as it had been practiced since Archimedes, Galileo, Newton, Faraday and Einstein. The other path led away from science as proposed by Mill, Pearson and Fisher. The path leading away from science is a research methodology, the aim of which was to *induce* empirical regularities. That inductive method is called 'empiricism'. Unlike science's proven track record in explaining, predicting and cumulating knowledge, the proponents of empiricism could not point to a record of successes to justify adopting its methodology. Instead of that justification, proponents of empiricism claim they are doing science. They are not.

D. Willer, P. Emanuelson, *The Logic of Theoretical Sociology*,
https://doi.org/10.1007/978-3-031-91709-7_1

Empiricists have no use for theory. Instead of theory, general knowledge is to be found by induction. Here is J. S. Mill's definition of induction:

> Induction, then, is that operation of the mind, by which we infer that what we know to be true in a particular case or cases, will be true in all cases that resemble the former in certain assignable respects. (1843/2012, p. 352)

By 1934, Popper (1934/1959), in his *Logic of Scientific Discovery*, had decisively demonstrated that the principle of induction cannot be justified. Later in this book, we show that there is no inductive method. Nevertheless, one year after Popper, Fisher published his 1935 *Design of Experiments*. In spite of Popper's proof, Fisher's design was inductive. It asserted that regularities found in the laboratory could, by induction, be fully general. They cannot. Empiricism is grounded in a fundamental logic error—*affirming the consequent*. It is true that, "If a relation is general, it can be found in the lab" but it is not true that "If a relation is found in the lab it is general". E*stablishing a relation in the laboratory is not evidence that the relation is general.* But, in spite of these faults, sociology followed the empiricist path.

In the decades following the middle of the twentieth century, it was becoming increasingly obvious to many sociologists that their field was not developing as a science. The teaching of research methods was unhelpful. As conventionally taught, research methods were part of the problem, not the solution. Progress in the successful sciences very frequently stemmed from the interaction of theory and experimental research. In sociology that interaction was notably absent. Instead, experimental research was categorized as "small group research," the scope of which was limited to groups with only a few members. Furthermore, experimental research was of limited use because it lacked 'external validity'. Exactly what kind of research had external validity was not addressed. Experimental research was shuffled to one side and the focus shifted to survey research. But survey research, unguided by theory, produces, at best, not generalizations, but inferences that are limited to the particulars of a population at a particular time and at a particular place. Nevertheless, as seen in later chapters of this book, developing theory in interaction with experiments has not been abandoned.

Chapter 3 of this book contains an apparent disjunction because it groups together the work of the sociologist, Robert Merton, with three philosophers of science, Hempel (1965), Popper (1959) and Kuhn

(1962). As just mentioned, in the decades following the middle of the twentieth century, it was becoming increasingly obvious to many sociologists that our field was not developing as a science. Merton offered a two-pronged solution that has proved to be remarkably prophetic. His two-fold solution is *middle range theory* and *consolidation of multiple middle range theories* to gain more detailed and broader scope explanations. Of the two, only the first was widely noted. As a result, many came to believe that Merton was proposing to narrow scope in order to increase a theory's precision. But there is not the slightest hint in Merton that narrowed scope would increase precision.

Merton was impeded in explaining what middle range theory would look like by the poverty of social theory at the time. For example, two of his examples, 'Role Theory' and 'Reference Group Theory' are not today seen as theories. Nevertheless, today there are active middle range theories. Status Characteristics Theory is a prominent example. As shown in Chap. 8, the scope of Status Characteristics Theory has expanded in a number of directions. Furthermore, consolidations in the 'Theory Toolbox' form are beginning to appear. Chapter 9 of this book gives examples of theory consolidation by Theory Toolbox.

The philosophers of science, Hempel and Popper, vehemently disagree on many issues. Insofar as the grounding of theory is concerned, Hempel is an inductivist and Popper is a deductivist. Nevertheless, the two agree that, in explanation, theory is applied deductively. Both propose that theory application is syllogistic: the major premise is a law from a theory; the minor premise is the particular facts, and the conclusion is the explanation. As seen in Chap. 3 this view is fraught with problems. Hempel and Popper each offer but a single example of deductive explanation. In presenting their example, however, both fail to cite the law that would form the major premise. Rather, the examples offered are trivial applications of the Stress = Strain law. As a result, both fail to make the case for explanation by deduction. Later in this book an alternative understanding of how theory is used in explanation is provided.

Kuhn is included because his influence on sociology at various times has been pernicious. Why it has been is easily seen. For Kuhn (1962) scientists form closed 'communities' that share a 'faith' in their 'paradigm', a faith to which they all 'adhere' (pp. 157–158). The authors of this book have practiced science throughout their academic careers and have yet to encounter closed scientific communities that share a faith in their paradigm. However, there would be no difficulty finding many closed religious communities that share faith. Kuhn is confused. Science is not religion.

Chapter 4 first reviews the replication controversy. Replication is seen as problematic because subsequent experiments frequently gave results distinct from otherwise identical previous experiments (Camerer et al., 2016; Open Science Collaboration, 2015). This conception of replication presumes induction. It presumes that induction can connect previous and current laboratory findings. As seen in Chap. 2, however, there is no inductive method. Therefore, there is no reason whatsoever to expect empiricist results to replicate.

Though theory receives no mention in either replication study, only theory can transcend the limits of the laboratory. This is easily shown by separating the experiments with no theory from those experiments that used at least some theory. The authors of this book carried out that separation and named experiments that used some theory hybrids. Hybrids all replicated. But all experiments with no theory failed to replicate. This is a distinction that foreshadows later chapters. For theory-driven experiments, replication is but a first step. More important is what could be called 'scope replication'. Scope replication goes beyond replication of a single design to replicate across a series of experiments to investigate the scope of the theory.

All scientific theory, including fully developed social theory, contains a viewpoint toward its phenomenon. In many cases the theory is built on that viewpoint. Since the viewpoint defines the nature of the world to which the theory is applied, it is called a 'worldview'. Chapter 5's investigations of worldviews assume that the way theory is structured is the same across the sciences. Therefore, there is much for sociologists to learn about the place of worldviews in theory by looking at worldviews of physical theories. The chapter begins with Aristotle's view of motion and how it was overthrown by the worldview shared by Galileo and Newton. Those worldviews conceptualize phenomena for explanation and prediction. The worldviews define the phenomenon that the theory is to explain and indicate how to frame that explanation. The authors find that the world Aristotle theorized was not the same world as theorized by Galileo and Newton.

Turning to social theory, the authors begin with Marx's worldview. For Marx, while history is made by thinking, acting humans, the ways that humans think and act are determined by preexisting social relations. Historical transitions occur when new social relations have grown within and emerge from the old. In that regard Marx's theory is deterministic. Importantly, determinism is not a fault neither here nor in any social theory. Social theory must see human activity as determined for, if it does not, it can say nothing about why people act the way that they do. Which is to say, anything claiming to be theory, if not determinist, is not theory.

How do new structures grow within the old order? Marx does not offer a general answer. His discussion of 'primitive accumulation', however, traces how the capitalist class structure grew out of late feudalism. Primitive accumulation, perhaps better named 'original accumulation', refers, not to the accumulation of capital, but to the accumulation of propertyless people who form the proletariat of the newly evolving capitalist social structure. Unlike many if not all earlier social structures, capitalism runs, not on coercive relations, but on exchange relations. Unlike coercive relations, the exchange relation itself is not a power relation. Yet power relations pervade capitalist social structures. How is this possible?

The answer is that capitalist social structures are filled with power structures. Power is produced structurally, not relationally. How power is produced by exchange structures is explained in Chap. 7 where Elementary Theory is introduced. Under capitalism there is always a surplus of workers over jobs. Yet being propertyless, the proletarians must work or die (Marx [1867] 1967, p. 596). The task for the sociologist is to discover further exchange structures that are the power structures of capitalist society. Inspired by Marx, the first part of Chap. 9 of this book traces how a new social structure grows within the old. Specifically, it traces how the early state grew out of earlier social structures.

Weber's goal, unlike Marx's, was to build theory for the newly emerging field of sociology while offering historical explanations, as in his *Protestant Ethic and the Spirit of Capitalism*. Unlike Marx, Weber does not see historical changes as determined by the economy alone. One example of that is in *The Protestant Ethic* (Weber, 1904/2002). Another is that Marx's primitive accumulation, as seen by Weber, was not produced through economic forces alone. Weber interpreted the development of the capitalist economy to be due in part to the impact of Elizabethan labor laws. Those draconian laws were understood by Weber to be evidence that the state contributes to economic change.

Weber in Chap. 1 of his *Economy and Society* was intentionally founding sociology as a science by formally introducing a technical vocabulary (Weber, 1918/1968). He introduced social action and social relationship with the obvious next step being social structure. Thus, social relationship could have been defined as pairs of mutually oriented social actions and social structure as composed of social relational parts. Defining more complex elements by simpler ones would have assured an internally consistent technical vocabulary. Unfortunately, Weber defined social relationship as a plurality of social actions and, instead of introducing social structure, took a tangent to introduce legitimacy instead. Why does Weber fail to be more systematic? Perhaps the fault did not lie in Weber. Unlike Galileo, Weber had no abstract geometry by means of which his technical terms could be represented and thus tied together.

Whereas worldviews are an essential part of scientific theory, not all worldviews lead toward testable theory. Homans' worldview is an example. Homans espoused what he called the 'covering law' idea of how scientific laws are applied. Though he does not mention Hempel, 'covering law' is certainly Hempel's term for the deductive application of theory. Homans' usage is much the same as Hempel's with one important exception. The 'propositions' that Homans proposes for his major premises are expressed as likelihoods. But nothing can be deduced from likelihoods. Therefore, Homans' deductive application of theory does not work.

Homans' worldview is of a psychologically driven social world. In that world there are no social structures. Moreover, there is nothing at the social level at all. In the years since Homans formed his worldview, many theories of social structure have been developed and tested. For example, the authors of this book compared the predictive precision of ten exchange theories, all of which predicted exchange ratios from structural conditions (Willer & Emanuelson, 2008). By contrast, Homans' rejection of social structure implies that all exchanges, being devoid of structural effects, would be the same. The ten theories varied in precision. However, all predicted far better than Homans'. All ten theories clearly falsify Homans.

Coleman's worldview is far better informed than Homans'. Coleman recognizes that social structures are an essential part of any sociological explanation. And a great strength of his approach is his understanding of economic theory. At the same time his reliance on economic theory is a shortcoming. In his view, sociology should adopt 'economic man' as its rational actor. Certainly, sociological theory can fruitfully use rational actors. However, it would be better to adopt game theory's strategic actor than economics' parametric actor. The terms 'strategic' and 'parametric' are explained in various contexts in this book.

Coleman expressed concern that social research was too far divorced from social theory. He saw sociological research as concerned, not with social structures, but with individuals and their attitudes. Were sociological research composed only of survey research, Coleman would be right. Certainly, from the 1950s through the 1970s survey research, with its focus on individuals and their attitudes, was so dominant that Coleman's complaint would be largely right. But by 1990, the date of publication of Coleman's *Foundations of Social Theory*, times were changing. By then a number of schools of thought concerned with social structures were connecting theory and experimental research. And none were concerned with individuals and their attitudes.

Chapter 6 reviews five contemporary theories and schools of thought, all of which are more formal than classical theories. Two of the five are game theory and network analysis. Both have strong mathematical grounding. Taken together the two contain most of the parts needed for a complete sociological theory. The parts include a social actor with values, beliefs and decision-making procedures and formulations for social structure. However, to our knowledge, game theory and network analysis have never been used together. Because each stands alone, both are incomplete theories. Furthermore, since the two have developed separately, there are no bridges that can be used to pull them together.

As seen repeatedly in Chap. 6, the great promise of the theories reviewed has gone unfulfilled because each is incomplete. If sociology is to be a science, broader scope, more exact theories must falsify narrower less precise theories. That selection process is a quality of a science, and not of any of the theories of which it is composed. The chapter concludes with a brief history of Power-Dependence Theory, a history that shows that sociology fails to select for scope or precision. Since it does not select, sociology is not functioning as a science.

Elementary Theory, introduced in Chap. 7, is given a chapter of its own because it is a complete sociological theory. It is complete in that it has social actors with (1) values, (2) beliefs and (3) decision-making procedures. Taken together the three make up what Marx called consciousness and Weber called meaning. The theory is completed by its diagramming procedure that is flexible enough to build up images of a wide variety of social structures. Peopling those diagrams with social actors generates dynamic models. Dynamic models are derivations from the theory that are testable. Once experimentally tested, explanations are generated, and predictions formulated.

'Complete social theory' does not mean theory to which nothing can be added. To the contrary, Elementary Theory is complete in that it has all the needed parts. Being complete, the expansion of its scope is straightforward. It began with a single structural power condition. Today seven more power conditions have been modeled and tested. Two modes of collective action have been formulated and both have been shown to countervail power and, in some cases, even to reverse it. All eight power conditions are introduced in Chap. 7. Taken as a whole our knowledge of social structures and their dynamics has cumulated through these investigations.

The authors believe that Elementary Theory is but one of an array of complete social theories that could be formulated and tested. The applications of Elementary Theory all occur within the scope of "Power Structures and Their Dynamics". The scope of sociology is far broader than the investigation of social power and, because it is, there are opportunities for wholly new theories to be developed. It is hoped that the explanation of the theoretic methods employed in building Elementary Theory can further theory building.

The structure of experiments is the focus of Chap. 8. The chapter begins with a review of an experiment published in the prestigious journal *Science*. The experiment is Fisher design and astonishingly trivial. That discussion is followed by a review of four experiments that are concerned in various ways with extending the scope of Status Characteristics Theory. Many extensions are long overdue—it particular the application of Status Characteristics Theory beyond the dyad.

In the final section of the chapter, we discuss how Elementary Theory experiments are designed. There it can be seen how theoretical models design the empirical replicas that test them. It would not be an exaggeration to say that the bulk of the work in building Elementary Theory experiments is at the level of theory. Once the theoretical model has been formulated, building its empirical replica for the lab is straightforward. A series of experiments have extended the scope of Elementary Theory and, arguably, produced cumulate knowledge about power structures.

Chapter 9, the final substantive chapter, takes up the problem of context complexity. The 'theory toolbox' is the proposed solution. The theory toolbox employed in that chapter contains only experimentally tested theories. Because they have been experimentally tested, the theories offer a particularly strong basis for historical explanation. Three explanations are offered: (1) the origin of the political state, (2) the transition from state property system of antiquity to feudalism and (3) the equilibrium achieved between the ancient Chinese and the status-lineage structures of the horse nomads.

For explanations, like those of Chap. 9, that apply lab tested theory outside the laboratory, context complexity can be a problem. Laboratory tested theory is purposefully simple while the world outside the lab is complex. Importantly, that is precisely the problem that the theory toolbox is designed to address. While each theory in the toolbox is simple, taken together they are complex in their implications.

The concluding chapter of this book is intended to tie together its parts into a systematic whole. Our goal in writing this book has been, first of all, to weed out the intellectual underbrush that stands between the prospective theorist and the goal of building, testing and applying theory. Because there is no inducive methodology, the one and only route to general knowledge in science is through theory. Given the central importance of theory, it is a surprise that there is no comprehensive source for how to construct theory and how to test it.

Look as you will, but exactly how to advance any science through theory is rarely if ever discussed. That the methods of empiricism are spelled out in detail is not helpful for they lead to a dead end. Nor is it helpful that philosophers of science are so parsimonious with examples. Even the best, Popper, tells the student little about the structure of theory and how it is formed. Sadly, even in the successful sciences where work with theory is ongoing, no clue is given as to how to build, test and apply theory. Why? From working with theory, the authors have learned that it is hard work to build theory. It is harder yet when there is no guide toward how it is built. It is hoped that this book can be that guide.

References

Camerer, C. F., Dreber, A., Forsell, E., Ho, T., Huber, J., Johannesson, M., Kirchler, M., Almenberg, J., Altmejd, A., Chan, T., Heikensten, E., Holzmeister, F., Imai, T., Isaksson, S., Nave, G., Pfeiffer, T., Razen, M., & Wu, H. (2016). Evaluating replicability of laboratory experiments in economics. *Science, 351*(6280), 1433–1436. https://doi.org/10.1126/science.aaf0918

Coleman, J. S. (1990). *Foundations of social theory.* The Belknap Press of Harvard University Press.

Fisher, R. A. (1935). *The design of experiments.* Oliver and Boyd.

Hempel, C. G. (1965). *Aspects of scientific explanation and other essays in the philosophy of science.* The Free Press.

Kuhn, T. S. (1962). *The structure of scientific revolutions.* The University of Chicago Press.

Marx, K. (1967). *Capital.* International Publishers. (Original work published 1867).
Mill, J. S. (2012). *A system of logic: Ratiocinative and inductive* (Vol. 1). Cambridge University Press. https://doi.org/10.1017/CBO9781139149839 (Original work published 1843).
Open Science Collaboration. (2015). Estimating the reproducibility of psychological science. *Science, 349*(6251), 1–8.
Popper, K. (1959). *The logic of scientific discovery.* Harper & Row. (Original work published 1934).
Weber, M. (1968). *Economy and society* (Vol. 1). University of California Press. (Original work published 1918).
Weber, M. (2002). *The Protestant ethic and the spirit of capitalism* (S. Kalberg, Trans.). Roxbury Publishing Company. (Original work published 1904).
Willer, D., & Emanuelson, P. (2008). Testing ten theories. *Journal of Mathematical Sociology, 32*, 165–203.

PART I

Empiricism and the Stagnation of Social Theory

CHAPTER 2

The Fork in the Path

Strangely, the path through which sociology acquired its *empiricist* research methods is not well known. That path is traced here. Empiricist methods are intended to discover fully general universals directly from particular research findings. Empiricist methods do not employ theory. Instead, it is said that empiricists employ *induction* to move from particulars to generalities.

From the mid-nineteenth to mid-twentieth centuries, a series of three English empiricists invented a new inductive research methodology. The first of the three was John Stewart Mill. His world view and methods, as proposed in his *System of Logic*, centered on his *canons of proof.* The canons required that the world be full of regular relations among events. Regardless of whether the world is as regular as Mill thought it to be, it did not prove to be regular in a way that could be captured by his canons.

The failure of Mill's methods led to a major shift in the empiricists' world view and a corresponding shift of method. The results of both shifts are found in the second of the English empiricists, Karl Pearson. Whereas Mill's methods relied upon finding perfect regularities, in Pearson's view, perfect regularities cannot be found because they do not exist. As explained in his *Grammar of Science*, contrary to Mill, the world is not composed of perfect regularities. Instead, the world is composed of irregularities of varying degree. His product-moment correlation was one part of Pearson's new methodology designed to find extent of regularity and irregularity.

D. Willer, P. Emanuelson, *The Logic of Theoretical Sociology*, https://doi.org/10.1007/978-3-031-91709-7_2

As a student of Pearson, the third English empiricist, Ronald Fisher shared some if not all of Pearson's world view. Of the three, Fisher was the most important to sociology and related social sciences because his research methods are still current. In his *Design of Experiments*, Fisher proposed means by which relations that are somewhat regular can be sorted out from others that are less regular and, thus, not counted as related. Based on statistical tests and Mill's canon of difference, Fisher proposed the experimental methodology that is dominant in social sciences today. Nor, as will be seen, has his influence been limited to experimental research. In sociology, his experimental method was adapted to the analytic survey.

The methodology introduced by the English empiricists allows, and even encourages, research without theory of any kind. As a result, it was the adoption of Fisher-design for experiments and surveys that separated theory and research in sociology. That separation has had a number of unfortunate consequences. As seen in this chapter, being devoid of theory, the products of Fisher-design experiments cannot explain or predict, nor can they produce cumulative knowledge. The failure of Fisher-design experiments to replicate is explained in Chap. 4.

John Stuart Mill

The key to understanding Mill's methods is to first grasp his understanding of the nature of the universe. Unlike empiricists today, in Mill's view, the "course of nature is uniform" (1843/2012, p. 371). It is uniform because "the universe is governed by general laws" (1843/2012, p. 371). Nevertheless, that uniformity may not be immediately obvious because "the uniformity of the course of nature is…itself a complex fact, compounded of all the separate uniformities" (1843/2012, p. 381). "The order of nature, as perceived at first glance, presents at every instance a chaos followed by another chaos. We must decompose each chaos into single facts" (1843/2012, pp. 437–438). That is to say, the task of science is to find the underlying universality by analysis—by cutting and separating the component parts of nature as found. The goal of this process is to uncover laws of nature where those laws are "the expression of the will of a superior; the superior in this instance being the Ruler of the Universe" (383–384). For Mill, the laws of nature are such because God created them. Beyond uniformity, the things we observe are *connected* to one another. "Every phenomenon is related in an uniform manner, to some phenomena that coexist with it, and to some that have preceded or will follow it" (1843/2012, p. 392).

It is by induction that Mill would establish regularities and find the connections among things.

> Induction, then, is that operation of the mind, by which we infer that what we know to be true in a particular case or cases, will be true in all cases which resemble the former in certain assignable respects. (1843/2012, p. 352)

And elsewhere, induction "may, then, be summarily defined as Generalization from Experience" (1843/2012, p. 370). Mill goes on to explain that the "all" in "all cases" does not mean merely all cases we have thus far observed, but all cases past, present, and to come. The aim of induction is to generalize from particulars, and the generalizations that result are *universal* truths. Mill's inductions are intended to be (as said in the subtitle of his *System of Logic*) the "*Methods of Scientific Investigation*".

However, Mill's inductions appear to have nothing to do with science or scientific laws. His examples, "fire burns" and "All men are mortal" are not the kind of general statements that occur in the sciences. Worse, there is reason to believe that, contrary to Mill, neither is a generalization from experience. It is undoubtedly true, as Mill asserts, that we believe that fire burns tomorrow because fire burns today, did yesterday and does in Cochin-China (1843/2012, p. 371). However, the authors of this book have seen films of people running through flaming hot coals who were not burned. Thus, the generalization, "fire burns" is false because it is not true in all cases. On the other hand, if fire is defined by its heat, we may intend "burns" to be a necessary part of the meaning of fire. If so, "fire burns" is definitional and not a generalization arrived at inductively. Nor can it be falsified. Turning to "All men are mortal", scholars of ancient Greece explain that, in Homer and more generally in the Greek myths, human heroes are men and the gods are not because gods are immortal. If so, "All men are mortal" is not based on a generalization from observation, but a religious belief.

Newton's *Principia* was published more than a century before Mill's *System of Knowledge*. Therefore, by the time *System* was written, many general formulations and derivations were available for Mill to employ. (See Chap. 5 for examples.) Nevertheless, very few examples are given, and when they are, their treatment by Mill is questionable at best. For example, Mill explained that the oscillations of a pendulum would soon come to stop due to the friction of the air. Placing the pendulum in a bell jar and evacuating the air around it allows the pendulum to oscillate longer: the better the vacuum, the longer the decay process. From this example Mill concluded that "the result was the proposition known as the first law of motion" (1843/2012, p. 476).

There is no doubt that the pendulum did oscillate longer as more and more air was evacuated. Nevertheless, Mill's conclusion is false on two grounds. First, Newton's first law of motion is, "*Every body continues in its state of rest, or of uniform motion in a right line, unless it is compelled to change that state by forces impressed upon it*" (Newton, 1686/1966a, p. 13). Pendulums oscillate. They do not continue in a right line. Therefore, Newton's first law does not apply.

Second, Newton's first law is not a generalization from experience as Mill would have it. To the contrary, it is a principle of Newtonian mechanics that forms a part of a great many derivations that explain and predict the motions of objects. If it were a generalization, unless it was carefully qualified, it would be false and the derivations employing it would be wrong. But they are not. Why did Mill make such an elementary error? It helps to see it from his worldview. In that world there are particulars and generalizations from particulars. There are no principles. Nor is there any place for mathematics in his worldview. In his *System of Logic*, no mathematics is found. This instance suggests that examples from science found in *System* are few, because Mill's worldview encompasses empiricism, not science.

Mill's canons were intended to be methods for finding generalizations by decomposing seeming chaos into separate uniformities. Of his five canons, his Method of Difference was the canon that Fisher developed into his experimental method.

Mill's Method of Difference states:

> *If an instance in which the phenomenon under investigation occurs, and an instance in which it does not occur, have every circumstance save one in common, that one occurring only in the former; the circumstance in which alone the two instances differ, is the effect, or cause, or a necessary part of the cause, of the phenomenon.* (1843/2012, p. 455)

Mill explains: "If the effect of A B C is a b c, and the effect of B C…[is]… b c, it is evident that the effect of A is a" (1843/2012, p. 454). Mill offers no examples of the Method of Difference. Nevertheless, he asserts that

Difference is the method of experimentation. It applies when all is the same but for one factor being studied. To be the same but for one factor, however, proved impossible to attain. The solution to the problem of exact similarity had to await Ronald Fisher and his randomizing between experimental and control groups.

Karl Pearson

The key to understanding Pearson's methods is to grasp his conception of the universe—his worldview—and how that universe is apprehended. For example, he asserts that there are no straight lines or flat planes in the universe (1892/2007, p. 198). Lines and planes are simplifications of what we perceive. They are most emphatically not abstract ideas. According to Pearson, "A scientific law is...the *rèsumè* or *brief expression* of the relationships and sequences of certain groups of these perceptions and conceptions and exist only when formulated by man" (1892/2007, p. 82).

And again. "There is more meaning in the statement that man gives laws to Nature than in its converse that Nature gives laws to man" (1892/2007, p. 87). Pearson frequently repeats that laws are summaries of sequences of sense impressions.

> We are thus to understand by a law in science, *i.e.* a "law of nature" a rèsumè in mental shorthand, which replaces for us a lengthy description of the sequences among our sense-impressions. (1892/2007, pp. 86–87)

Written laws are produced by sequences of sense impressions and nothing more. What is written is nothing more than a convenient summary of those sense impressions.

Scientific laws are products of the human mind, not laws legislated by God. Because laws only describe relations and sequences (1892/2007, p. 99), it follows that laws "never explain the routine of our perceptions" (p. 99). They merely describe that routine. Because laws are descriptions, finding laws does not introduce necessity (113). Furthermore, "That the future will be like our experience of the past is the sole condition under which we can predict what is about to happen and so guide out conduct" (138). This justification of induction is repeated. "The overwhelming probability drawn from past experience in favor of *all* sequences repeating themselves at once embraces the new sequence" (141). Here Pearson's position is deterministic.

But elsewhere he is not a determinist. It is a "fundamental truth that nothing in the universe repeats itself" (170). Furthermore,

> No phenomena are causal; all phenomena are contingent, and the problem before us is to measure the degree of this continency, which we have seen lies between the zero of independence and the unity of causation. (1892/2007, p. 174)

Since 'correct' understanding of the universe sees all phenomena as correlated, never causal (p. 177), the use of equal signs in the statement of scientific laws is a mistake. There is nothing in the universe to justify it. Because equal signs in laws are not justified, the coefficients of his product-moment correlations should replace equal signs. Pearson modestly called his replacement of cause with contingency "epoch-making" (p. 165).

Since they are not justified by experience, Pearson proposed removing the equal signs from laws like D = VT (distance equals velocity times time) and F = MA (force equals mass times acceleration) and substituting one or another correlation coefficient in their place. That substitution was justified because it improves the fit between perceptions and statements about perceptions. But substituting his coefficient for the equal signs is a mistake. For example, D = VT does not mean that the measures of distance, velocity and time are approximately equal, but that V = D/T, that velocity *is* distance divided by time. And F = MA *means that* acceleration of any given mass is proportionate to the force applied to it. It also means that, for any given force, mass and acceleration vary inversely... and similarly. Looking to scientific measurement, the physical dimensions of both sides of F = MA are mass, distance and time squared. They are identical as they must be for the law is an identity. Since it is an identity, it is true by definition and cannot be arrived at by induction or any other form of generalization.

Why did Pearson make such elementary errors? Why did he believe that the equal signs in scientific laws could only be justified by perfect regularities? The answer is that this error is not an isolated instance. Rather it reflects his worldview. The error is a straightforward consequence of his view that laws are shorthand descriptions of sense impressions. Also consistent with his view is the assertion, at the beginning of *Grammar*: "Nobody believes now that science *explains* anything; we all look upon it as a shorthand description, as an economy of thought" (xi [italics original]). Were science nothing but a shorthand record of sense impressions as he maintains, it could not explain anything. It could not because the instance to be explained is already contained in its generalization. (See Chap. 3.) To the contrary, science does explain, and it is theory that gives the explanation.

There is good reason to doubt whether Pearson's view of the universe is internally consistent. His thinking is, at first, deterministic and then indeterministic. It is indeterministic when he asserts that regularities are only relatively regular. It is deterministic when he asserts that those relatively regularities are themselves regular. Beyond odd, the two assertions verge on mutual contradiction. They make no sense together. Nor can it be seen how the two conditions, the "irregular regularity" that he proposes, and the "regular irregularity" that he condemns, could be empirically distinguished.

Given Pearson's misunderstanding of science, it is perhaps inevitable that much of *Grammar* is pseudoscience. For example, Pearson 'corrects' Newton, by rewriting the laws of motion (311–353). Pearson's rewrite, that we will not repeat here, is much more complex than the original, eliminates Newton's elegance, and, worse, eliminates the laws' capability of functioning as principles. As Newton wrote somewhere, "Truth is ever found in simplicity", a good rule for the theorist and one that Pearson did not follow. Had Pearson's revision been generally adopted, the whole of Newtonian mechanics would have tumbled down. But it was not, and it did not.

Late in the *Grammar,* Pearson argues for the necessity of the idea of ether. Ether was the presumed medium through which electromagnetic waves were transmitted. The *Grammar* was first published in 1892, whereas the Michelson and Morley experiment disproving ether was published five years earlier in 1887 (Michelson & Morley, 1887). Soon after that publication their results were widely known in the scientific community. Though the Michelson and Morley experiment failed to find the expected effect of ether, that result seems not to have influenced Pearson. He continued to champion ether as a necessary concept of science. Other odd ideas include Pearson's proof that Kepler's laws of planetary motion do not contain necessity. His proof is this. "I might go around the ellipse a hundred times in the same manner and then stop or go off in an entirely different path" (135). Of course, Kepler's laws apply to planets, not to the whims of Pearson.

Today Pearson's views on the nature of science are of historical interest only. Why then does mainstream social science methodology still use

Pearson's correlation coefficient as a measure of relatedness? It seems that the intellectual foundations of Pearson's thought are now dead, but his methodological technique, the correlation coefficient, zombielike, shambles on.

Ronald Fisher

As a student of Pearson, Fisher may have inherited his world view from his teacher. By focusing on a particular kind of experimental design, however, Fisher is clear that his focus is on generalization from particulars, which is to say on induction. His goal in *The Design of Experiments* was "to exhibit the principal designs which have been found most successful in that field of experimentation, namely agriculture, in which questions of design have been most thoroughly studied" (1935, p. 9).

That most thorough study in the field of agriculture to which he referred was his own. Prior to joining the faculty of University College London, he studied agricultural data at Rothamsted Experimental Station for 14 years. Students of the history of science will be astonished at Fisher's offhand rejection of the long tradition of theory-driven experimentation as practiced by Archimedes (1897), Galileo (1954) and Newton (1966a, 1966b).

From experiences at Rothamsted, he proposed to explain "the principles which are common to all experimentation" (1935, p. 9). How are all experiments designed? In its simplest form, Fisher was proposing an updated version of Mill's method of difference.

Fisher's contributions to experimental design solved two fundamental problems left by Mill: (1) to assure that method of difference comparisons across experimental and control groupings were between groupings that differ in only one regard, and (2) to resolve when an experiment indicates a difference and when it does not. Having solved both problems by introducing control by randomization, Fisher-design experiments could reach conclusions that Mill-designs could not.

Regarding Fisher's first contribution, it is no simple task to produce only a single difference between experimental and control groupings. Every experimental setting contains an uncountable number of things that could affect the experimental outcome in unknown ways: the education of experimental subjects, their gender, the current weather, the effect of the color of the lab's walls on the mood of the experimental subjects, and so forth. Fisher's solution equated experimental groupings by randomly assigning the units composing those groupings to either the experimental or control condition. The units being assigned could be, as in Fisher's own research, a variety of seeds or squares of land marked off within fields (1935, pp. 46ff). Alternatively, as in research in sociology, the units could be persons. Random assignment does not produce groupings identical in every regard. Instead, it produces groupings that are similar on the average.

As will be remembered, Mill intended that his methods be applied until an absolute difference was found. It was only an absolute difference that assured the researcher that a relation had been discovered. Experience has shown, however, that absolute differences cannot be found by Mill's methods. If the method of difference is to be employed, another criterion is needed.

Separating results that indicate a difference from results that do not is the task of the test of significance. For example, units having been randomly assigned to two groupings, one of the two is subjected to an experimental treatment. A "null" hypothesis is proposed: that the two groupings differ only by chance—due to variations introduced in random assignment. The null hypothesis is required "because it must supply the basis of the 'problem of distribution' of which the test of significance is the solution" (1935, p. 16). With random assignment, a test for the likelihood of difference can be run—for example the *t*-test or *f*-test—where the estimates of error are "derived from the data" (33). If the null hypothesis is rejected, the experimenter may decide to assert its opposite—that a difference was found due to the treatment. According to Fisher, "(e)very experiment may be said to exist only in order to give the facts a chance of disproving the null hypothesis" (1935, p. 16). Given the Fisher-design, it remains only to choose the likelihood at which to reject the null hypotheses. He suggests .05.

There is no doubt that Fisher's solutions to Mill's two problems were substantial advances. Oddly, at no point in his *The Design of Experiments* did he single out either of the two as advances. Instead, his emphasis was squarely on induction.

> We may at once admit that any inference from the particular to the general must be attended with some degree of uncertainty, but this is not the same as to admit that such inference cannot be absolutely rigorous…The mere fact that inductive inferences are uncertain cannot, therefore, be accepted as precluding perfectly rigorous and unequivocal inference. (1935, p. 4)

The process of leaving behind the particulars of the laboratory to attain generalizations about the world was "rigorously uncertain". But to Fisher that uncertainty could not be avoided. "Inductive inference is the only process known to us by which essentially new knowledge comes into the world" (1935, p. 7).

Fisher's goal was induction from particulars of the laboratory to general knowledge about the world. Yet the methodology to attain that goal is strangely absent. Where in Fisher-design is any technique of induction to be found? Said somewhat differently, what part of Fisher-design supports induction beyond the particulars of the laboratory? Exactly how is fully general knowledge produced from particular findings? In fact, there is nothing in Fisher's design that supports induction. A search of the whole of *The Design of Experiments* will find no technique that produces inductive inferences. Is there a technique for inductive inference? If there is, none of the three empiricists discussed here offers it.

Establishing a relation in the laboratory is not evidence that the relation is general. Yet Fisher, and Mill as well, implicitly assume that finding a relation is proof of generality. Here the word "implicitly" is essential. In our reading we have not found either Mill or Fisher to explicitly claim, each by their own criterion, that a found relation is necessarily fully general. Nevertheless, Mill, Fisher and undoubtedly other empiricists have made the fundamental error of *affirming the consequent*. It is true that, "If a relation is general, it will be found in the lab" but it is not true that "If a relation is found in the lab it is general".

Nor should finding that error come as a surprise. One year before the publication of Fisher's *The Design of Experiments*, Karl Popper's *The Logic of Scientific Discovery* was published. In that work Popper pointed out that 1) science cannot be based on induction because 2) induction cannot be justified. To justify induction "we should have to employ inductive inferences; and to justify these we should have to assume an inductive principle of a higher order; and so on" (1934/1959, pp. 28ff.). To justify induction fails because it leads to "infinite regress" (1934/1959, pp. 28ff). To Popper it can be added that there is no induction methodology in experiments of the Fisher-design. Fisher-design itself does not produce general knowledge.

Nevertheless, some experiments seemingly employing Fisher-design do result in general knowledge. How is this possible in the absence of induction? The answer lies in hybrid design. Hybrid design gains generality by combining elements of Fisher-design with general theory. Hybrid design is discussed in detail in Chap. 4. As shown there, insofar as replication is concerned, pure Fisher-designs are ineffective, but hybrid designs do replicate. Conversely, the reproducibility of research results implies that the experiments are hybrid designs—they use theory to gain generality. They are not pure Fisher-designs.

How do hybrid designs work? Every one of the hybrid designs reviewed in Chap. 4 uses game theory as a part of the design. Game theory *is* theory and its introduction in the hybrid design experiments was the essential element that allowed those experiments to transcend the particulars of the lab—allowed them to be general knowledge. General knowledge replicates, particulars do not. (See Chap. 4.)

Something more should be said here about how hybrid designs work. The contribution of the Fisher-design component of the hybrid is to create experiment and control groupings by random assignment to 1) assure similarity between groups prior to the treatment and 2) to test statistically for the likelihood of difference between the groupings. Had the design stopped there it would have been a pure Fisher-design and its results would be particular to the lab—but it did not. The 'treatment' in some hybrid designs was game theoretic. In one example, the conditions of the control grouping satisfied game theoretic conditions while the conditions of those in the experimental grouping did not. Game theory predicted different acts for the two groupings, and the predictions were satisfied. It was game theory that rendered the predictions general.

Looking back at Fisher's own research as he presented it in *The Design of Experiments*, some results appear general and, when they do, quite a bit of biological theory can be found embedded in them. The theory of inheritance developed late in the nineteenth century by Gregor Mendel and chromosome theory by Boveri-Sutton were integrated by Thomas Morgan in 1915 to form the core of classic genetics (Gayon, 2016). These developments were carried forward by Fisher himself in his 1930 book, *The General Theory of Natural Selection*. Contrary to Fisher, it was not induction, but biological theory that gave generality to the experiments he discussed.

At a number of points in *The Design of Experiments*, Fisher asserted that his experimental design is common to all the sciences. He claimed further that all experiments exist only to reject a null hypothesis. These are pernicious claims, but it appears that they are believed today. For example,

> In summary, the method of difference and the principle of random assignment provide the logical and statistical foundation for contemporary experimental design and analysis. Mill and Fisher left in their wake a powerful set of analytical and statistical tools that have become the method of understanding causality in science. (Thye, 2014, p. 67)

To the contrary, the practice of experimental research in the sciences did not await Fisher's 1935 publication of *The Design of Experiments*. The tradition of scientific experimentation is found as early as 232 BCE in *The Works of Archimedes*, again in 1636 during the Renaissance in Galileo's *Two New Sciences* and, of course, in a great many publications since that date. Later in this book much will be said about how scientific experiments are structured. For example, in the sciences null hypotheses are unknown. At this point it is sufficient to say that theoretical sciences do not now and never previously have employed Fisher-design. That scientific research results are reproducible is evidence that they are general knowledge.

The Analytic Social Survey and the Problem of 'Correlated Biases'

The analytic social survey was introduced to sociology in the 1940s. The logical foundation for it was the Fisher experiment. By the 1950s experimental design in sociology meant Fisher-design. Shown in Fig. 2.1 is Stouffer's (1950) idea of the ideal experiment. Its origin in Fisher is seen in its experimental and control groups, neither of which are found in theory-driven experiments. (See Chap. 7 of this book.) It depicts a method of difference design in which only the experimental group is subject to a 'treatment'. Measures on both groupings are taken before and after the treatment and differences are compared.

Stouffer asserts that the difference between experiment design and survey designs is whether the design includes information on all or only some of the four boxes of the Figure. Experiments are said to contain all four. By contrast, a social survey may have only one grouping that is studied at two points in time, thus contained in just two of the four boxes. Alternatively, a survey may have two boxes, such as rural – urban or male – female samples, which are compared to explain, for example, voting preference. Below we explain why, contrary to Stouffer, the random assignment of Fisher-design experiment and the random selection of the analytic social survey are not the same.

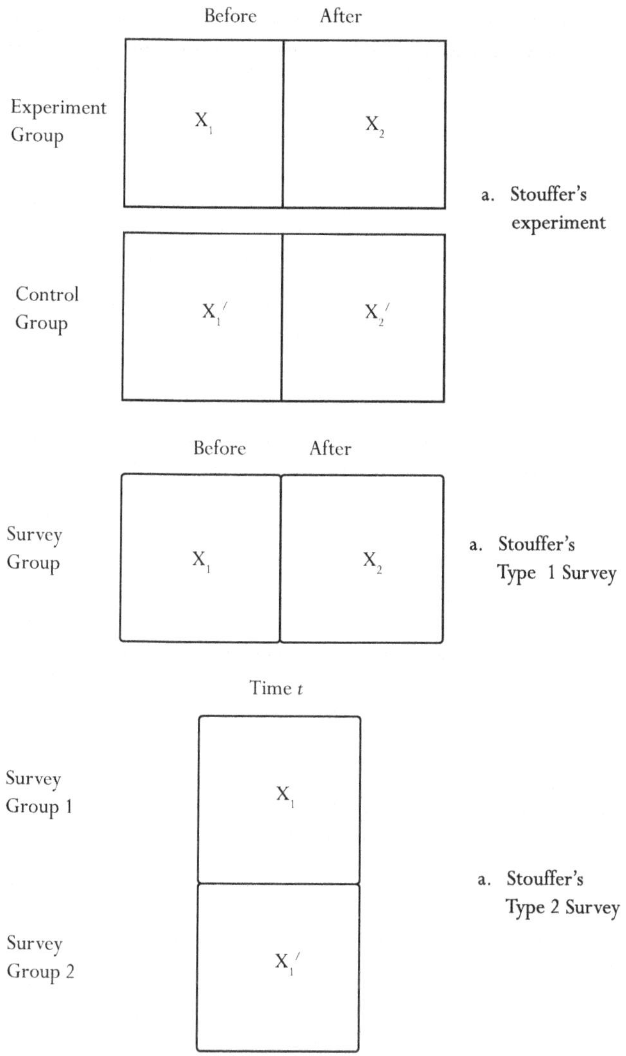

Fig. 2.1 Stouffer's design for an ideal experiment

Nevertheless, contemporary to Stouffer, a new kind of analysis of survey data called 'elaboration' had been introduced by Kendall and Lazarsfeld (1950) and others. Elaboration works like this. Assume that a survey shows that those who live in cities vote more frequently Democratic than their rural neighbors. Is this difference due to something about place of residence influencing voting, or is it due to the better education of urban dwellers? Could it be due to the greater propensity of urbans to join clubs? Or is it due to the greater religiosity of the country dwellers? These and other questions can be answered by parceling out subsamples and testing for significance.

For Stouffer the design of the experiment and the design of the survey differed only in the number of boxes—that is to say, the amount of information—available for analysis. That characterization is wrong. Whereas both designs owe much to Fisher, the designs of survey and experiment are fundamentally different. Only in the experiment are units such as experimental subjects randomly *assigned* to experimental and control groupings. Only through random assignment do we have reason to believe that the two groupings differ only by chance prior to the running of the experiment. And only then does the null hypothesis justify the distributions such that they can be statistically tested. Because of random assignment, Fisher-design experiments cannot have what has come to be called 'correlated biases'. Furthermore, given random assignment, Stouffer's pretest boxes are redundant. Fisher-design requires only the two of Stouffer's boxes on the right.

Randomness in the survey is random *selection* of a sample from a population. Random selection justifies inferences from the sample to the population from which the sample was drawn. But random selection does not produce experimental and control groups that differ only by the treatment that one of the two receives.

As a result, the categories such as rural—urban or working class—middle class that survey analysis uses are not at all like the categories of experiment and control. As Selvin (1957) explained:

> The two groups in our example may differ in many ways in addition to the differences produced by sampling and by the rural-urban difference whose effect we are trying to measure. Thus, urban residents are, on the average, more educated, have higher incomes, smaller families, more memberships in formal organizations, and so on. That is, these characteristics are correlated

> with the variable of urban-rural residence, which is why they are called "correlated biases". (p. 522)

As Selvin points out: "Sociologists cannot randomize except in laboratory studies of small groups" (1957, p. 521). He concludes that, "Since there is no way of knowing, in general, the sizes of these correlated biases and their directions, there is no point in asking for the probability that the observed differences could have been produced by random errors" (1957, p. 527).

Selvin did not long go unchallenged. A year after Selvin's paper appeared, McGinnis (1958) offered a critique. He began by granting that correlated biases pose problems for the survey researcher (412), but then argued at length that the problems could be resolved. In any case sociologists had no choice but to use test of significance in survey research. What would happen if those tests were disallowed? "One consequence is clear: "most social researchers would soon be out of business" (412). That is to say, "the business slump would follow swiftly and inevitably" (413). Whether the prospect of a business slump is a sound criterion on which to judge a method in science we leave to the judgement of the reader. Sounding much like Pearson, McGillis made the following assertion. "There is no such thing as a completely general relationship which is independent of population, time, and space" (412). It would seem that, for McGinnis, Newton's F = MA, Einstein's E = MC^2 and many other scientific laws do not exist.

Conclusion: The Path Taken

By the mid-twentieth century, sociology faced a fork in the path of research methodology—as did those parts of other social sciences that shared its methodology. Sociology could have followed the path of science, the path of Archimedes, Galileo, Newton, Faraday, Einstein and others. It could have worked to develop theories, test them in the laboratory, and apply them to explain and predict social events. That choice would have wed sociology to the, by then, long standing tradition of theoretically driven research. We do not claim that all research needs to be theory-driven. Policy research, for example, might have little connection to developing theory, or it might well benefit from new understandings flowing from scientific developments. In either case, policy research stands on its own feet and does not need further justification.

Today we know that science was not the path taken. Instead, sociology adopted the methodology proposed by the three English empiricists just reviewed. Theory as method was cast aside in favor of finding relative regularities. Induction from particulars to universals was the goal. Yet no one, including Mill, Pearson and Fisher, offered a method for induction. Let it be clear. *To have found a regularity or relative regularity in the lab or through another form of investigation is not evidence for its universality. To attain universality from observational data would require the use of a method of induction. Thus far, no one has proposed that method.* Furthermore, a generalization that is not universal is merely a statement about an enumerable set. It cannot explain because generalizations simply sum up particulars and nothing more. Theory, not induction, is the proven path to general knowledge.

One point is abundantly clear from this chapter's survey of the sources of empiricism as now practiced in sociology. Not one of its authors, not Mill, not Pearson and not Fisher, had the slightest understanding of how theoretic science works. The path of empiricism started with Mill's unworkable method of difference, traced through Pearson's destructive proposal to replace equals signs with correlation coefficients and cumulated in Fisher's falsehood that all experiments exist only to reject a null hypothesis. It is a path that traces error after error.

The methodology introduced by the English empiricists allows, and even encourages, research without theory. It calls for inductive inferences but offers no methodology for induction. The results of empiricist research do not explain, do not predict and do not cumulate. As seen in Chap. 4, recent research shows that empiricist experiments fail to replicate.

Nevertheless, empiricism teaches this. Since no one has discovered a methodology for induction, scientific laws and scientific theory cannot be inductive generalizations. This is not a question of whether the world is as regular as Mill thought it to be or as irregularly regular as Pearson and Fisher believed. No one can say anything definitive about the regularity or irregularity of events in this world. What can be said is that scientific theory imposes regularity on the world and empiricism does not.

Can general statements concerning either regularities or probable regularities explain? Regularities do not explain nor do statements about them explain. As seen in this chapter, there is no inductive method that spans from a countable set to a corresponding universal set. Therefore, generalizations are not universal. They are statements about particulars. It follows that statements about new observations join the list of statements about particulars that compose the generalization. Just as the new observation statement cannot explain itself, it cannot be explained by some or all of the particular statements that compose the generalization. Therefore, generalizations do not explain. They also do not predict.

The wrong path was taken. It is high time to take the other path. It is high time to build theory and do science.

References

Archimedes. (1897). *The works of Archimedes* (T. L. Heath, Ed.). Dover Publications. (Original work published ca. 232 B.C.E.).

Fisher, R. A. (1935). *The design of experiments.* Oliver and Boyd.

Galileo, G. (1954). *Dialogues concerning two new sciences* (H. Crew & A. Desalvio, Trans.). Dover Press. (Original work published 1636).

Gayon. J. (2016). From Mendel to epigenetics: History of genetics. Comptes Rendus. *Biologies, 339*(7–8), 225–230. https://doi.org/10.1016/j.crvi.2016.05.009. https://comptes-rendus.academie-sciences.fr/biologies/articles/10.1016/j.crvi.2016.05.009/

Kendall, P. L., & Lazarsfeld, P. F. (1950). Problems of survey analysis. In R. K. Merton & P. F. Lazarsfeld (Eds.), *Continuities in social research: Studies in the scope and method of the 'American Soldier'* (pp. 133–196). Free Press.

McGinnis, R. (1958). Randomization and inference in sociological research. *American Sociological Review, 23*(4), 408–414. https://doi.org/10.2307/2088804

Michelson, A. A., & Morley, E. W. (1887). On the relative motion of the Earth and the luminiferous ether. *American Journal of Science, 34*, 333–345. https://doi.org/10.2475/ajs.s3-34.203.333

Mill, J. S. (2012). *A system of logic: Ratiocinative and inductive.* Cambridge University Press. https://doi.org/10.1017/CBO9781139149839 (Original work published 1843).

Newton, I. (1966a). *Mathematical principles of natural philosophy and his system of the world: The motion of bodies* (A. Motte, Trans., 1729, Vol. 1). University of California Press. (Original work published 1686).

Newton, I. (1966b). *Mathematical principles of natural philosophy and his system of the world: The system of the world* (A. Motte, Trans., 1729, Vol. 2). University of California Press. (Original work published 1687).

Pearson, K. (2007). *The grammar of science.* Cosimo Classics. (Original work published 1892).

Popper, K. (1959). *The logic of scientific discovery.* Harper & Row. (Original work published 1934).

Selvin, H. C. (1957). A critique of tests of significance in survey research. *American Sociological Review, 22*(5), 519–527. https://doi.org/10.2307/2089475
Stouffer, S. A. (1950). Some observations on study design. *American Journal of Sociology, 55*, 355–361. https://doi.org/10.1086/220558
Thye, S. R. (2014). Logical and philosophical foundations of experimental research in the social sciences. In M. Webster Jr. & J. Sell (Eds.), *Laboratory experiments in the social sciences* (2nd ed., pp. 53–82). Elsevier Academic Press. https://doi.org/10.1016/B978-0-12-404681-8.00003-0

CHAPTER 3

Social Science and Its Myths

By the middle of the twentieth century, it was becoming increasingly obvious that sociology was not developing as a science. What should be done? Textbooks on research methods were of little help. When influenced by Mill, Pearson and Fisher, as many were, they were part of the problem, not the solution. People in sociology with experience in more successful sciences suspected that theory was the solution. But the social theory of the era did not appear to be at all like theory in sciences like physics or chemistry. Historically, some sociological theory, if theory it was, was strongly influenced by Darwinian evolutionary biology. Too often that influence did not aid in explaining social phenomena. Instead, it morphed into an ideological Social Darwinism that, instead of explaining social phenomena, focused on victim blaming—blaming the poor for the faults of a wealthy and overindulgent leisure class (Hofstadter, 1992).

As discussed further in Chap. 5 of this book, a number of influential scholars in sociology saw economics as more advanced than sociology. And economists were then—and still are—generous in expressing the judgment that their science, unlike sociology, builds on a widely accepted core of knowledge (Fehr & Gintis, 2007). Could sociologists learn how to build a progressive science by emulating economics? Two issues stood in the way of that emulation. First, inferring from economic theory's grounding on rational man, it was thought that social theory should be grounded on normative man. While rational man optimized rational decisions, social

D. Willer, P. Emanuelson, *The Logic of Theoretical Sociology*, https://doi.org/10.1007/978-3-031-91709-7_3

man would act on internalized norms. This seemed a plausible suggestion. Since no one then proved able to build testable theory grounded on normative man, however, this idea died.

Second, at midcentury, few in economics saw the need to test mainstream economic theory. Possibly surprising to outsiders, much of economic theory was taken as true, at times even necessarily true. Economics was what V. Smith called an "*a priori* science" (1989, p. 151). If inconvertibly true, theory need not be tested. Perhaps this is the reason why experimental research in economics lagged behind experimental research in other presumably less developed social sciences like sociology and social psychology. As long as little need was seen to test it, economic theory took on the qualities of a great frozen mass of truths. Not until economic theory and experimental research were harnessed together, as they are increasingly today, would economics become a developing science.

Sociology has a distinguished classical theory tradition of its own. Particularly notable are the works of Marx and Weber. Their contributions appear as great Gothic cathedrals and it is by no means obvious what part of their work is the fundamental skeleton upon which one could build a science. To the authors' knowledge, Marx did not comment on whether there was a place for experimental testing of any part of his theoretic. Weber did comment briefly on linking meaningful interpretation of action to experimentation. "Unfortunately this type of verification is feasible with relative accuracy only in the few very special cases susceptible of psychological experimentation" (1918/1968, p. 10). Since classical theory was not seen at midcentury as the way to kick-start a developing science of sociology, we will return to it in Chap. 5 under orienting perspectives.

One further point needs to be addressed. There is a longstanding view in sociology that laboratory research, because it investigates groupings in which the number of individuals is relatively few, is necessarily small group research. As will be seen in later chapters of this book, that view is simply wrong. Whether group size in the laboratory limits applications outside is wholly a question of the capabilities of the theory being applied.

Whereas there are a great many methods books in sociology, there are very few in the most successful sciences. A successful science explains, predicts and cumulates knowledge. Obviously, a successful science has successful theoretic methods. Seeking clues toward those methods, sociologists fruitlessly searched their campus libraries. What they did not find were methodology texts in physics and chemistry. What they did find were works in the philosophy of science. At that time, the dominant school of philosophy of science was logical positivism, a school of thought that was influenced by the English empiricists reviewed in the previous chapter. A central concern of logical positivism was to distinguish between science and religion. For a time, it was thought that, unlike religion, the concepts and propositions of science could all be reduced to observational statements—a position that Hempel (1965) called the "narrower thesis of empiricism". While that thesis ruled out religion, an analysis previously offered by Carnap (1936, 1937) showed that it also ruled theory out of science. In any case, resolving the narrower thesis did not offer the guidance sought by sociologists.

During the mid-twentieth century, a number of books on theory construction written by sociologists began to appear. They included works by Zetterberg (1954), Stinchcombe (1968), Blalock (1969), Dubin (1969), Mullins (1971), Reynolds (1971), Wallace (1971), Gibbs (1972) and Hage (1972). That list may give the impression that there was a coordinated effort in sociology to develop theoretic methods. In fact, there was no coordinated effort (Willer, 1996). The books just listed did not build on one another, nor did the later ones analyze the ideas proposed by the earlier ones. The authors of the theory construction books did not meet periodically to share ideas. Since none of those authors had the opportunity to work with a formal, testable social theory nor had any seen that kind of theory, none of the authors were actually qualified to write about the development of formal, testable social theory. These works, like standard methods texts in sociology, offered no effective guidance toward developing sociology as a theoretic science. By contrast, the authors of this book have multiple decades of work building, testing and applying formal scientific theory. (See Chaps. 7, 8 and 9.) That experience grounds the guidance offered here toward developing sociology as a theoretical science.

This chapter looks at what might be learned about how theories are formulated and applied from a selection of writings from around the mid-twentieth century. The discussion begins with Merton's view of the future of social theory, and then moves to three philosophers of science: Hempel, Popper and Kuhn. For Hempel the focus is on his deductive-nomological view of theory application for explanation. The discussion of Popper reviews his contention that theory is applied deductively and is never the result of induction. His thoughts on rational choice as a basis for sociological model building is reviewed. The conclusion critically evaluates Kuhn's view of the nature of science.

Robert Merton

Merton made many contributions to sociology. Perhaps the most notable was his idea of middle range theories that are:

> theories that lie between the minor but necessary working hypotheses that evolve in abundance during day-to-day research and the all-inclusive systematic efforts to develop a unified theory that will explain all of the observed uniformities of social behavior, social organization and social change. (1957, p. 39)

While Merton does not take up what the structure of middle range theories should be, he gives 'reference group theory' and the 'theory of role sets' as examples. Regarding the latter he says that it begins with a concept and imagery that generate theoretical problems. Middle range theories are not the musings of a master theorist nor are they simply empirical generalizations. That is to say, they are not the sort of thing produced by Talcott Parsons or Paul Lazarsfeld.

Are middle range theories testable and, if so, how? It seems that, neither reference group theory nor role set theory were ever subject to anything like a test. In fact, it is doubtful that today either reference group or role set would be counted as theories. Instead, both would be seen as orienting perspectives. Though Merton is silent regarding the need to test middle range theory, his repeated calls for rigor strongly suggest that he would be supportive of testing any theory and most particularly middle range theory.

Whether called middle range theory or not, in the years since Merton, a great many theories of modest but significant scope have been put forward and subject to rigorous experimental test. To take but one example, Bornstein and associates have theorized the relation between external conflict and internal interest situations. Experiments by Bornstein and associates (Bornstein, 2003; Bornstein et al., 1990, 2002) demonstrate that threats of intergroup conflict interact with group payoffs to reduce free riding and increase group cohesion. Further studies are supportive. Puurtinen and Mappes (2009) found that intergroup competition resolves collective action problems by increasing group cooperation. Further studies have shown that external threats increase support for leaders (Gavrilets & Fortunto, 2014; Van Vugt et al., 2008).

Theory relating external conflict to internal solidarity offers a particularly striking example in which the rigor of lab-tested theory applies forcibly in the field. Pursuing war to increase solidarity and support of leaders is a particularly nefarious yet frequently pursued policy. As documented by R. Willer (2004), the President of the U.S., George W. Bush, when facing rapidly declining support, demanded higher taxes and increased powers and subsequently pursued war justified by imaginary threats of mass destruction by Iraq. By attacking Ukraine, V. Putin, whose hold on power in Russia has been weakening, is pursuing a particularly bloody war that is used to justify political suppression within Russia. These examples clearly demonstrate that the implications of laboratory research are not limited to the study of small groups.

Merton proposed that theories of the middle range should be *consolidated* to gain broader scope. By consolidation he means that middle range theories should somehow be linked together in application. Consolidation is repeatedly taken up (1957, p. 51, 53, 64), suggesting that it is an important part of the middle range proposal. How is consolidation to be done? Merton quoted the physicist Feynman to the effect that "today our theories of physics, the laws of physics, are a multitude of different parts and pieces that do not fit together very well" (Quoted by Merton, 1957, p. 48). Given the lack of consolidation in physics, how could consolidation be possible in sociology? In Chap. 8 of this book, the authors propose, not a consolidation, but an amalgamation of theories. Founded on the image of a "theory toolbox" the amalgamation is an array of theories that together cover the phenomena to be the explained. The explanation consists of pulling one and then another theory from the toolbox and applying them in sequence to the broader phenomena. Examples are given in Chap. 9.

Perhaps Merton's call for middle range theories and their consolidation was, not a call, but a prediction of future developments in sociology. That prediction is dead on. Today sociology is rich with experimentally tested formally stated theories of intermediate scope. Because it is, at least part of the field is progressive science.

Carl Hempel

According to Hempel, the purpose of scientific research is to discover regularities in order to establish general laws that can be used for explanation, prediction and postdiction (1965, p. 173). What are laws and theories? Laws are true statements (1965, p. 265) of universal form (p. 266) and of nonlimited scope (p. 265). Laws are true in that they are tested. They are of universal form as in "All metals conduct electricity" and "All apples in this basket are red". And they are of nonlimited scope (p. 267). Only the statement about metals, and not the statement about apples, is of nonlimited scope and thus only it could be a law.

A central part of Hempel's philosophy of science is his "deductive nomological" scheme of how science explains, predicts and postdicts. The scheme is analogous to the syllogism with its minor premise, major premise and conclusion. Hempel diagrams it this way (1965, p. 174):

$$\frac{\begin{array}{c} C_1, C_2 \ldots C_K \\ L_1, L_2 \ldots L_K \end{array}}{E}$$

The *Cs* are "statements of particular occurrences", the *Ls* are general laws, and the *E* is "the sentence stating whatever is being explained, predicted or postdicted" (p. 174). *E* follows deductively from the *Cs* and *Ls* (174). While the '*C*' in Hempel's deductive scheme strongly suggests 'cause' and '*E*' 'effect', Hempel himself does not call them cause and effect. It seems that the *Ls* are abstract laws. If so the *Ls* and the *Cs* and *E* are not at the same level of abstraction. To link them for deduction requires a series of "*interpretative sentences*" (p. 184).

Having distinguished the empirical level from abstract laws, Hempel provides a critique of "The Theoretician's Dilemma" (p. 185). The dilemma is this: If abstract laws and theory serve their purpose, they establish general connections among observational phenomena. If they do they can be dispensed with by replacing them with statements that directly link observations (p. 186). If so, theory is superfluous. As Hempel points out, however, the abstract terms currently in use in the sciences cannot be fully defined in terms of observationals. Furthermore, the abstractions of science contribute to its "heuristic fertility" whereas observables do not. Therefore, the Theoretician's Dilemma "rests on false premises" (1965, p. 222).

For the sociologist seeking an understanding of how science works, however, resolving issues like the Theoreticians Dilemma are only of tangential interest. Of more central interest is the question, "How are explanations, predictions and postdictions formed?" Hempel is undoubtedly right that "Most explanations offered in history or sociology, however, fail to include an explicit statement of the general regularities they presuppose" (1965, p. 236). Given that critique, the reader eagerly awaits Hempel's own examples of his deductive nomological scheme. Nevertheless, in his paper, "The Function of General Laws in History" (chapter 9 of 1965), Hempel gives not a single example of a scientific explanation of a social or historical event.

He does, however, give an explanation of the "cracking of an automotive radiator on a cold night" (1965, p. 232). The essential parts of his explanation are as follows. The car was left in the street overnight when temperature dropped to 25°. "The bursting pressure of the radiator material is so and so much" (p. 232). When water freezes "the pressure of the mass of water increases with decreasing temperature" (p. 232). He adds that the explanation would have to include "a quantitative law concerning the change of pressure of water as a function of its temperature and volume" (p. 232).

Hempel's 'explanation' is faulty on two grounds. First, the law he fails to include is the stress-strain law. It states that stress = strain within the limits of perfect elasticity. The elastic strain limit of the radiator would be found by experimental test. Second, it is not water that provides the stress that cracks the radiator, but the pressure produced by the expansion of water freezing into ice. The stress is determined by the amount that water expands as it freezes. That quantity is also found by experimental test.

The simple elegance of Hempel's deductive nomological scheme is brought into question by the parsimony of his examples and his mistakes when an example is given. Hempel refers to laws as general regularities, general hypotheses and universal empirical hypotheses. Could these be the regularities that J.S. Mill failed to find? Hempel also offered another scheme that he called "inductive-statistical" (1965, pp. 275ff), where the general statement refers, not to phenomena that are perfectly related, but to phenomena that are related by a likelihood. Could these be the relative regularities of Pearson and Fisher? Hempel answers neither question. While not questioning the philosophical value of Hempel's work, it offers little help to the sociologist seeking to learn how theory is formed and applied.

Karl Popper

Popper sharply differs from Hempel on the issue of how laws and theories are grounded. Hempel is an inductivist. Laws and theories are discovered by generalization from particulars (Hempel, 1965, pp. 3–98). Popper disagrees. To justify induction,

> we should have to employ inductive inferences; and to justify these we should have to assume an inductive principle of a higher order; and so on. Thus, the attempt to base the principle of induction on experience breaks down, since it must lead to infinite regress. (1959, p. 29)

In the place of induction, Popper proposes a "*deductive method of testing*" (1959, p. 30). Theory is falsified only if a reproducible effect falsifies it (p. 86). Theories are not falsified one at a time. Testing is a competitive process. In falsification: "We chose the theory which best holds its own in competition with other theories" (p. 108). Furthermore, scientific theories are intersubjectively tested (p. 44). That theories are falsifiable follows from the "logical form of universal statements" (1959, p. 41). Universal statements "are never derivable from singular statements, but can be contradicted by singular statements" (p. 41). Popper labels as myth the idea that science begins with observations and experiments and proceeds to theory (p. 279). He includes thoughts from Einstein in his appendix. There Einstein asserts "that theory cannot be fabricated out of the results of observation, but that it can only be invented" (p. 458).

What then is the relation between a newly tested theory that has survived falsification and the old theory that was falsified? For Popper, the old theory often still applies as "kind of a limiting case" (p. 252). That is to say, over at least some part of the scope of the new theory, the predictions of the old theory apply, at least as approximations. Said somewhat differently, the new theory "contain*s*" the old at least as an approximation (p. 276). While Popper gives no example, the relation between Newton's and Einstein's theories is a case in point. For velocities significantly short of the speed of light, Newton's theory gives many predictions similar to Einstein's. (See below.)

Popper is straightforward in driving home his idea of falsification. Scientific hypotheses do not all have the same value. A theory that makes a greater number of falsifiable predictions is superior to one making fewer. Here, in our view, Popper is not quite right because all quantitative theories are capable of making an infinite number of testable predictions. It seems to us that the issue cannot be number of predictions, but size of scope of predictions. For Popper, theories are either falsified or not. They are not falsified with a probability. His position, contrary to Pearson and Fisher, is that "it is useless and misleading to employ the concept of probability in connection with scientific hypotheses" (p. 316).

Popper sees prediction as but another aspect of the use of laws in explanation (1957, p. 114). As he asserts, "there is no great difference between explanation, prediction and testing" (p. 123). What constitutes a satisfactory explanation? For explanation "testable and falsifiable universal laws and initial conditions" are necessary (Popper, 1972, p. 193). Though Popper's terminology is not the same as Hempel's, his view of explanation and Hempel's are effectively the same. For Hempel's C_1, C_2, C_3...Popper uses "initial conditions" and for L_1, L_2, L_3...Popper uses "universal laws". There are important similarities between Hempel's deductive nomological scheme and Popper's view of explanation and prediction. As quoted above, however, Popper would not accept Hempel's "inductive statistical" scheme as explanatory.

For Popper a theory is either false or not. Falsification is a state. What if a test falsifies a theory? What should the theorist do? Is finding a theory false a dead end? Not necessarily. Perhaps falsification should be thought of as a process. The process begins with the first test of a theory. If that test falsifies, the theory is revised and tested again. Tests and revisions continue until a test does not falsify. The process is complete because theories that are not falsified can be used for explanation and prediction. This process is taken up later in the book.

Popper is as parsimonious as Hempel with examples of explanations. He gives but one, the breaking of a thread. If a weight of two pounds is placed on a thread with tensile strength of one pound, it will break.

> Whenever a thread is loaded with a weight exceeding that which characterizes the tensile strength of the thread, then it will break; a statement which has the character of a universal law of nature. (p. 60)

It is perhaps ironic that Popper and Hempel selected something breaking as an example of scientific explanation. Hempel's example was flawed in many ways. Popper's example is better handled; however, he also failed to cite a law. The universal law of nature is not about a thread. Instead, as in Hempel, the law in question is the stress = strain law. Like Hempel, Popper gives far too few examples. Nevertheless, it seems to us that Popper's grasp of how science works is substantially better than Hempel's.

Particularly meritorious is Popper's treatment of "uniformities of nature" (p. 253). The question of whether nature is uniform is not, according to Popper, subject to test. He prefers the "invariance of laws" (p. 253). There is an important difference between the two positions, particularly for the neophyte seeking to formulate a theory. The authors of this book have invented and tested a theory. That experience has led us to believe that the uniformities given by the theory are as much a product of the theory itself as they are a product of the phenomena being theorized. The world may well be complex, a blooming, buzzing confusion, but theory can be simple and can impose order.

How are theories invented? Popper draws a line between inventing a theory, an act not subject to logical analysis, and the testing of theory that is subject to logical analysis. "Accordingly I shall distinguish sharply between the process of conceiving a new idea, and the methods and results of examining it logically" (p. 31).

While there are no rules for inventing theory, testing theory is a process that is subject to logical analysis. Between inventing and testing, however, there is an important issue that remains unaddressed in Popper. *What are the elements from which theories are composed and how should they be structured?* For Popper theories are composed of statements, but what constitutes a statement remains undefined. Not found in Popper is an explication of what theory is and how its parts are structured. For the sociologist seeking to understand how to build theory, Popper is no guide. He is not alone. Had philosophers of science been the methodologists of science, the lack of examples would be unforgivable. But they are not. Philosophers of science are philosophers and are under no such obligation.

Nevertheless, Popper offers a number of valuable insights into science. For example, contrary to Fisher, he says that the "experiment is planned action in which every step is guided by theory" (p. 280). Popper is right. Later chapters show how, in science, theory designs experiments. Here is a second example. Whereas some assert that sociology's slow development has been due to the very great complexity of social phenomena, Popper disagrees. He says that there is a

> widely held prejudice that social situations are more complex than physical ones .... we are liable to compare what should not be compared; I mean on

> the one hand concrete social situations and on the other hand artificially insulated experimental situations. (Popper, 1957, p. 129)

A third example gives insight to his thinking on rationality and model building. Beyond being no more complex, Popper asserts further that social situations can be simpler to model than physical ones because of the unique role of rationality in model building. As he points out, by employing rationality, relatively simple models for human actions and interactions can be constructed (1957, p. 130). Demonstrations of that simplicity are found in later chapters of this book.

Beyond Popper, there is evidence that models employing rationality are simply better models. Chapter 4 addresses why some experiments replicate and others do not. Research reported there finds that experiments originally published in *Science* and *Nature* that employed rationality in the form of game theory replicated whereas those that did not employ rationality did not replicate. So, employing rationality is both a route toward simpler and also more effective models. Later chapters will show how rational actor models are built and how they are deployed.

Thomas Kuhn

Here is Thomas Kuhn's view of science. In science a 'community' shares a 'faith' in their 'paradigm', a faith to which they all 'adhere' (1962, pp. 10ff). Paradigms have 'textbooks' through which the orthodoxy of 'normal science' is communicated to the masses. The work of normal science is 'puzzle solving', an activity to which 'adherents' are 'passionately devoted' (pp. 35–37). The terms in single quotation marks are Kuhn's. That terminology implies that science is a kind of religion. It is a religion in which the orthodoxy of 'normal science' periodically changes. That change, called a 'scientific revolution', may or may not be progressive (p. 162).

How does Kuhn think science works? To find the answer look closely at any example he offers. Here is what he says about Einstein's general theory.

> As we shall see, ... there are seldom many areas in which scientific theory, particularly if it is cast in a predominantly mathematical form, can be directly compared with nature. No more than three such areas are even yet accessible to Einstein's general theory of relativity. (p. 26)

This is confusing. Does Kuhn mean that Einstein's theory has only ever been applied three times? Or does he mean that Einstein's theory has been applied to only three kinds of events?. Each meaning will be discussed.

Many years ago the older of the authors of this book worked on Purdue University's synchrotron, a particle accelerator. The synchrotron produces high energy electrons by accelerating them nearer and nearer to the speed of light. As the speed of light is approached the electrons gained mass and that mass gain resulted in the very high energies needed for the experiments being run. The mass gain, predicted by Einstein's theory, was experienced, not three times, but routinely, many times a day. Therefore, this meaning of Kuhn's is wrong. Now consider the second of Kuhn's possible meanings.

In 1919 Einstein wrote a letter to the *London Times*. In that letter he pointed out that, in many applications, predictions from his theory and Newton's were effectively identical, so much so that differentiating the two is difficult. Because the scope of his theory is broader, Einstein was able to point to three circumstances that differentiated: (1) revolution of the ellipses of planetary orbits, (2) gravitational curving of light rays and (3) red shift of radiation due to gravitation. Of these the first two had been confirmed at the time of Einstein's writing and the third has been confirmed since. In addition, his General Theory predicts gravitational waves, time contraction, the big bang, black holes and is the basis of the whole of modern cosmology. Furthermore, Einstein's theory "contains" (Popper's term) Newton's. Since all of Newton's theory is contained in the scope of Einstein's, it is also supported by all evidence supportive of Newton's theory. Thus, contrary to Kuhn, there is nothing "equivocal" (p. 26) about the support of Einstein's theory.

Reading Kuhn, one is left with a feeling of vagueness. This feeling has at least three sources. First, Kuhn never defines terms. Second, he invents new terms: neologisms with indeterminate meanings that are employed instead of already well-established terms. For example, 'paradigm' is substituted for 'theory'. Third, Kuhn, like Hempel and Popper, is parsimonious with examples. In *Revolutions* many scientific developments are mentioned, but he comments further on only one, Einstein's theory focusing on its support. As just seen, Kuhn is wrong about its support.

Is it true that Kuhn sees science as a kind of religion? His terminology suggests that he does. Another test is the issue of progress. To many it is perfectly obvious that science progresses, but religions do not. But if Kuhn does see science as a kind of religion, he must raise doubts about whether sciences progress. Here is how he does that.

> With respect to normal science…the problem of progress lies simply in the eye of the beholder. Scientific progress is not different in kind from progress in other fields, but the absence at most times of competing schools that question each other's aims and standards makes the progress of the normal scientific community far easier to see. (p. 162)

Kuhn goes on to say that the impression of progress is also due to the insulation of the scientific

community "from the laity" (p. 163) and its "narrow and rigid education" (p. 165). He concludes with this question. "Why should scientific communities be able to reach a firm consensus unattainable to other fields?" (p. 172). Kuhn gives no answer. The answer is that science employs objective criteria as seen in comments by Einstein on his theory above. The Times statement by Einstein adds following about his theory.

> The chief attraction of the theory lies in its logical completeness. If a single one of the conclusions drawn from it proves wrong, it must be given up; to modify it without destroying the whole structure seems to be impossible. (Einstein, 1919)

It is difficult to imagine a statement more in the spirit of openness to the testing of scientific theory and more contrary to a dogmatic rejection of the progress by Kuhn.

A decade after *Revolutions*, Kuhn took up the issue of what are "the characteristics of a good scientific theory?" (1974). Here is a paraphrase of his list.

1. A theory should be accurate.
2. A theory should be consistent internally and with other accepted theories.
3. A theory should have broad scope.
4. A theory should be simple and bring order to its phenomenon.
5. A theory should disclose new phenomena.

Kuhn's list is meritorious with one exception. Regarding the second part of his second point, certainly no theorist is obliged to compare their theory to other theories outside its scope. Though Kuhn's list is meritorious, what he did with it is something else. For him, the list does not assure selection of better theories because the choice between theories is a function of both "objective and subjective criteria" (p. 359). While for him the subjective criteria are more important, it should be pointed out that science is unique in minimizing the subjective component of decisions. One good example is Einstein's comments on his own theory above. Because subjectivity is minimized, science progresses.

For proof of the importance of subjectivity, Kuhn points to the competition between the theories of Copernicus and Ptolemy. As Kuhn points out, Copernicus' system was no more accurate than Ptolemy's (p. 357). Nor is that surprising because most of the data on which Copernicus based his theory was Ptolemy's (Koestler, 1964, p. 200). As Koestler put it, Copernicus "was not particularly fond of star-gazing" (1964, p. 123). Regarding simplicity, Copernicus noted that Ptolemy's system required 50 epicycles while his own required only 34. However, a revised version of the Ptolemaic system, contemporary to Copernicus, had only 30 epicycles (1964, p. 192). By Kuhn's criteria for good theory, the two theories were equal in accuracy and simplicity. In fact, neither theory displaced the other. That neither did is evidence that Kuhn's objective criteria work. But it is not evidence supportive of any subjective component.

The sociologist hoping to build scientific theory will not be helped by Kuhn. To the contrary, in spite of the excellence of his objective criteria for theory, he otherwise contributes only confusion about what scientific theory is and how it works.

Conclusion

Merton put forward two ideas, both of which are as important today as they were more than half a century ago. He called for building middle range theories and for consolidating them into ever broader formulations. The idea of theories of the middle range is still current. The idea that they should be somehow consolidated has fallen through the cracks.

There are at least two ways that broader scope can be gained. First, the scope of a theory can be broadened. Every year any number of excellent investigations, including many experimental tests of social theories, are published. Frequently, they extend previous work, but, when they do, the extensions are scattered across several journals. That they are scattered is particularly egregious. Cumulation is occurring now in sociology, but it is largely if not wholly invisible. Obviously, the norms of science are not working well in sociology. Nevertheless, much of the fault lies with sociology's journals. New papers that expand the scope of previously tested theory should be given priority in the journal that published the initial statement of the theory. But frequently they are not. (For an example see the discussion of power-dependence theory in Chap. 6.) As a result, the impact of programs where a theory's scope is being extended is lost and with it its increased significance.

Secondly, following Merton, combining theories is another way to gain broader scope. How is that to be done? Looking ahead, it seems that social phenomena are most often composed of many and varied parts. At least at this stage in the development of our science, it seems wrong to try to cover that complex whole with a single theory. To try to do so would almost assuredly be as vacuous as Parson's attempt. Far better to cover the parts of the phenomena with many theories, the applications of which can be systematically articulated. This is the thinking behind the 'theory toolbox' of Chap. 9.

The significance of much current research that tests theory will remain hidden until explanations of macrostructures are constructed out of the smaller theories now at hand. It is for reasons such as these that we see Merton's call to be as important today as it was when first written.

Hempel, Popper and Kuhn, as philosophers of science, knew full well that their field is not a methodology of science. And they were right. For the sociologist seeking insights into theory building, particularly disappointing is the very small number of examples offered. Perhaps the only point of similarity among Hempel, Popper and Kuhn is that, in their major works, each was satisfied to discuss at any length only a single example. Nevertheless, the well is not entirely dry. There is something to be learned from the three and also from the work of other philosophers of science, particularly Stephen Toulmin (1953).

References

Blalock, H. M., Jr. (1969). *Theory construction: From verbal to mathematical formulations*. Prentice Hall.

Bornstein, G. (2003). Intergroup conflict: Individual, group, and collective interests. *Personality and Social Psychology Review, 7*(2), 139–145.

Bornstein, G., Erev, I., & Rosen, O. (1990). Intergroup competition as a structural solution to social dilemmas. *Social Behavior, 5*, 247–260.

Bornstein, G., Gneezy, U., & Nagel, R. (2002). The effect of intergroup competition on intragroup coordination: An experimental study. *Games and Economic Behavior, 4*, 1–25.
Carnap, R. (1936). Testability and meaning (part I). *Philosophy of Science, 3*(4), 419–471. https://doi.org/10.1086/286432
Carnap, R. (1937). Testability and meaning - continued. *Philosophy of Science, 4*(1), 1–40. https://doi.org/10.1086/286443
Dubin, R. (1969). *Theory building*. The Free Press.
Einstein, A. (1919, November 28). What is the theory of relativity? [Letter]. *London Times*. https://germanhistorydocs.ghi-dc.org/pdf/eng/EDU_Einstein_ENGLISH.pdf
Fehr, E., & Gintis, H. (2007). Human motivation and social cooperation: Experimental and analytical foundations. *Annual Review of Sociology, 33*, 43–64.
Gavrilets, S., & Fortunto, L. (2014). A solution to the collective action problem in between-group conflict and within-group inequality. *Nature Communications, 26*, 1–11.
Gibbs, J. (1972). *Sociological theory construction*. Dryden Press.
Hage, J. (1972). *Techniques and problems of theory construction in sociology*. J. Wiley & Sons Ltd.
Hempel, C. G. (1965). *Aspects of scientific explanation and other essays in the philosophy of science*. The Free Press.
Hofstadter, R. (1992). *Social Darwinism in American thought*. Beacon Press.
Koestler, A. (1964). *The act of creation*. Hutchinson.
Kuhn, T. S. (1962). *The structure of scientific revolutions*. The University of Chicago Press.
Kuhn, T. S. (1974). Second thoughts on paradigms. In F. Suppe (Ed.), *The structure of scientific theories* (pp. 459–482). University of Illinois Press.
Merton, R. K. (1957). *Social theory and social structures*. Free Press. (Original work published 1949).
Mullins, N. C. (1971). *The art of theory: Construction and use*. Harper & Row.
Popper, K. (1959). *The logic of scientific discovery*. Harper & Row. (Original work published 1934).
Popper, K. R. (1957). *The poverty of historicism*. Routledge.
Popper, K. R. (1972). *Objective knowledge: An evolutionary approach* (Revised ed.). Clarendon Press.
Puurtinen, M., & Mappes, T. (2009). Between-group competition and human cooperation. *Proceedings of the Royal Society, 276*, 355–360.
Reynolds, P. D. (1971). *A primer in theory construction*. Taylor & Francis.
Smith, V. (1989). Theory, experiment, and economics. *Journal of Economic Perspectives, 3*(1), 151–169.
Stinchcombe, A. L. (1968). *Constructing social theories*. Harcourt, Brace & World, Inc.

Toulmin, S. (1953). *The philosophy of science: An introduction*. Hutchinson's University Library.
Van Vugt, M., Hogan, R., & Kaiser, R. (2008). Leadership, followership, and evolution. *American Psychologist, 63*(3), 182–196.
Wallace, W. L. (1971). *The logic of science in sociology*. Aldine Atherton.
Weber, M. (1968). *Economy and society*. University of California Press. (Original work published 1918).
Willer, D. (1996). The prominence of formal theory in sociology. *Sociological Forum, 11*, 319–331.
Willer, R. (2004). The effects of government-issued terror warnings on presidential approval ratings. *Current Research in Social Psychology, 10*, 1–12.
Zetterberg, H. L. (1954). *On theory and verification in sociology*. Bedminster Press.

CHAPTER 4

The Replication Problem

The recently reported failure to replicate experimental results in psychology and economics has sparked concerns across the social sciences (Anvari & Lakens, 2018). For psychology, the Open Science Collaboration (2015) found that only 36% of attempted replications succeeded, while Klein et al. (2018) found replication success for 54%. For economics, Camerer et al. (2016) found successful replication for only 61%. Of equal concern, effect sizes of replicants were smaller than for initial studies (Resnick, 2018; Klein et al., 2018).

To solve the replication problem, Starbuck (2016) suggested replacing tests of significance with variance explained, while Killen (2005) proposed a new estimate that he asserts captures the replicability of results (Killen, 2005). Alternatively, Benjamin et al. (2017) suggested keeping significance tests, but moving the standard for the level of significance from 0.05 to 0.005 to reduce the number of findings due to chance. Silberzahn and Uhlmann (2015) pointed out how choices at each step of an analysis biases outcomes and suggested 'crowdsourcing' analyses as a solution. (In crowdsourcing, a project is made public so that many different individuals can contribute.)

Others have pointed to 'perverse' career incentives that produce questionable methodological practices (Moody et al., 2022). Ravetz (1971) argued that science 'degenerates' when incentives reward the pursuit of grants and prestige over scientific integrity. 'Publish or perish' incentives

D. Willer, P. Emanuelson, *The Logic of Theoretical Sociology*,
https://doi.org/10.1007/978-3-031-91709-7_4

contribute to the file-drawer effect (Rosenthal, 1979; Franco et al., 2014; Ferguson & Heene, 2012; Gerber & Malhotra, 2008), causing publication bias (Earl & Trifimow, 2015). In a survey of 1500 scientists, Baker (2016) found that more than half of respondents cited pressure to publish, selective reporting, and low statistical power as contributors to failure to replicate. Casey et al. (2012) suggested that investigators should register study designs and hypotheses before running experiments. Do replications fail because of corrupt methodological practices such as HARKing (Hypothesizing After Results are Known) (Starbuck, 2016)? If any of those practices were abandoned, would the problem be resolved?

Alternatively, it has been suggested that human behavior is too complex to be captured scientifically (Earl & Trifimow, 2015; Lieberson & Lynn, 2002). However, the belief that social phenomena are more complex than physical phenomena is not universal. According to Popper (1957), there are good reasons to believe that social science is less complicated than physics (p. 130).

Perhaps the replication problem is an illusion because applying theory in all sciences requires inexplicit "auxiliary assumptions" (Earl & Trifimow, 2015). Looking at physics in the late 1960s, others could not then replicate Joseph Weber's claim to have detected gravitational waves. And, Pons and Fleischmann's purported observation of cold fusion in 1989 failed to replicate and is now widely rejected (Steinle, 2016, pp. 52–536). Despite failures to replicate in physics, no one has yet suggested physics has a replication *problem* of the same order as that found by the Open Science Collaboration, Klein et al. and Camerer et al.

Looking elsewhere for the source of this difference, we suggest that the methodologies of experimental investigations in physics and sociology are different. A clue that they differ is the absence of the word *theory* in the Open Science Collaboration (2015) and Camerer et al. (2016) studies. Making no mention of *theory* in either replication study is consistent with conducting experiments without using theory, a practice so widespread in sociology and some other social sciences that it is the norm. By contrast, Franklin (2018, p. 1) stated that in physics, most replication experiments "examine different phenomena that bear on the same theory or hypothesis".

Muthukrishna and Henrich (2019) suggest that theory solves the replication problem. But what is theory? Turner (1993, p. 7) pointed to "violent dissension over what theory is, what it should be, or what it can be". Turner rightly claims that theory has no fixed meaning in sociology. It follows that any attempt to separate studies that test hypotheses grounded in theory and those that are not will be controversial. That difficulty is avoided by not using the source of a hypothesis to differentiate theory-grounded experiments from those that are not. Instead, the focus here is on whether theory *designed* the experiment.

As seen in Chap. 2, Fisher-design experiments seek regularities. Thus, the measure of their reproducibility is whether findings can be regularly reproduced. For theory-designed experiments the measure of reproducibility is quite different. Here experiment results are not compared to experiment results but to a theory's predictions. This form of replication may or may not repeat a previous study. Replication may retest one model or test various models across a theory's scope. Testing across scope is not possible for Fisher-type experiments because, having no theory, they have no scope to explore. Therefore, Fisher-design experiments replicate only if an exact copy of an original experiment design gives the same results.

Replication across a theory's scope is not unlike Earl and Trifimow's (2015) "conceptual replication", the aim of which was "to see whether a given phenomenon...might obtain across a range of variable conditions". This type of replication is as old as scientific experimentation. It began in antiquity with Archimedes' (232 BCE/1897) study of the laws of levers and was taken up again during the Enlightenment by Galileo's (1665/1954) study of motion and Newton's (1686/1966a, 1687/1966b) mechanics. Indeed, theory-driven experimentation was long established when Fisher-design experiments arrived in 1935.

This chapter first shows why there is no reason to expect Fisher-design experiments to replicate. Then, turning to theory-designed experiments, a brief explanation of how theory *designs* experiments is offered. (A more detailed explanation of how theory designs experiments is given in Chap. 7.) Next, hybrid experiments that use parts from Theory-design and Fisher-design are explained. A review of Camerer et al.'s study shows that the hybrid experiments they tested were universally successful at replicating, while the Fisher-designed studies were rarely successful. Finally, replication across a theory's scope is discussed and why theory-design experiments *produce* successful replication is explained.

Fisher Design and the Problem of Induction

As seen in Chap. 2, Fisher-design experiments rely upon induction to extend research results beyond the limits of a single study. Induction is necessary for reproducibility. But no methodology for inductive generalization was offered by Fisher or his predecessors, nor has such been offered since. That assertion is based on two factors. First, as discussed in Chap. 2, Ronald Fisher's assertions to the contrary, nowhere in his *Design of Experiments* did he propose a methodology for induction. Second and more generally, if there were a methodology of inductive generalization, it would not be invisible. It would be well known—but it is not.

Why would a methodology of inductive generalization necessarily be well known? It would be well known because a form of inference superficially like induction is practiced every day in sociology and its methodology is widely known. We refer to inference from sample to population. As long as the sample is selected by some form of randomization, by applying probability theory, inference from sample to population is straightforward. Nevertheless, inference from sample to population is not inductive generalization nor does it give general knowledge. It is knowledge about particulars—as for example, knowledge about a particular population of people. The point is this. If there were an inductive method that produced fully general knowledge that method would be much more valuable than inference from sample to population. Therefore, that inductive methodology would be better known today than is inference from sample to population. But it is not because there is no such method.

Since there is no inductive method, inducive inference in Fisher-designs cannot produce general knowledge. Therefore, but by chance, Fisher-design experiments cannot replicate. It follows that there is only one method for attaining fully general knowledge: it is through application of theory. Popper might say that the only method for attaining fully general knowledge is through ***deduction*** from theory. Popper is wrong. Applying theory is not deductive. Exactly how theory is applied is explained with examples in the following chapters.

Theory-Designed Experiments

The replication of a theory-designed experiment succeeds if the experiment's outcome agrees with the theory's predictions (Walker & Willer, 2006, 2007; Willer & Walker, 2007). To start a replication, one or more models are derived from a theory. A model is the theory's phenomena in abstract form. The model is dynamic, aligning events into a process linking initial to final conditions. The process is the theory's prediction. When multiple models are derived, each model designs an experiment's replica or replicas, where replicas are concrete representations of theoretical models. Said differently, each replica serves as the design of an experiment. To conduct an experiment, derive a model, build a replica, set initial conditions, then observe dynamics and final conditions. For replication, minimally a correspondence between a model and its replica must occur.

When theory designs an experiment, the steps taken do not differ between the natural and social sciences. Take, for example, Meselson and Stahl's (1958) experiment to test alternate processes of DNA replication. Shortly after Watson and Crick introduced the double-helix structure of DNA, scientists proposed three competing models for DNA replication: called 'conservative', 'semi-conservative', and 'dispersive'. As shown in Fig. 4.1, each model for replication consists of the original DNA strand, as proposed by Watson and Crick (1953), and the predicted daughter strands after DNA replication. The model's mechanism is an enzyme-catalyzed polymerization process that transforms free nucleotides into new strands of DNA. That mechanism makes each model dynamic, with initial and final conditions predicting the creation of new DNA strands from available compatible materials. Their replica was a petri dish containing *E. coli* bacteria in a nutrient broth containing nitrogen, which bacteria can use to synthesize new DNA. Meselson and Stahl collected DNA for each new generation of bacteria.

Because Meselson and Stahl tested competing predictions, they needed to examine DNA post-replication. They began growing *E. coli* in a 'heavy' nitrogen isotope solution, and bacteria used the materials to create new generations of bacteria. Once they produced a generation where all the nitrogenous bases in the bacteria's DNA were labeled 'heavy' (Generation 0), the bacteria were switched to the broth using the 'lighter' isotope (Generations 1 and beyond). Generation 1 comprised the experiment's initial condition. Applying the model's dynamics, all new DNA must incorporate the new, 'lighter' isotope, and the configuration of that integration is the final condition. Meselson and Stahl's measure of their final condition used a process called density gradient centrifugation, which separates molecules such as DNA into bands that form a density gradient. Each model predicts a unique density gradient by the third generation of DNA replication. The test compared the end condition, the density

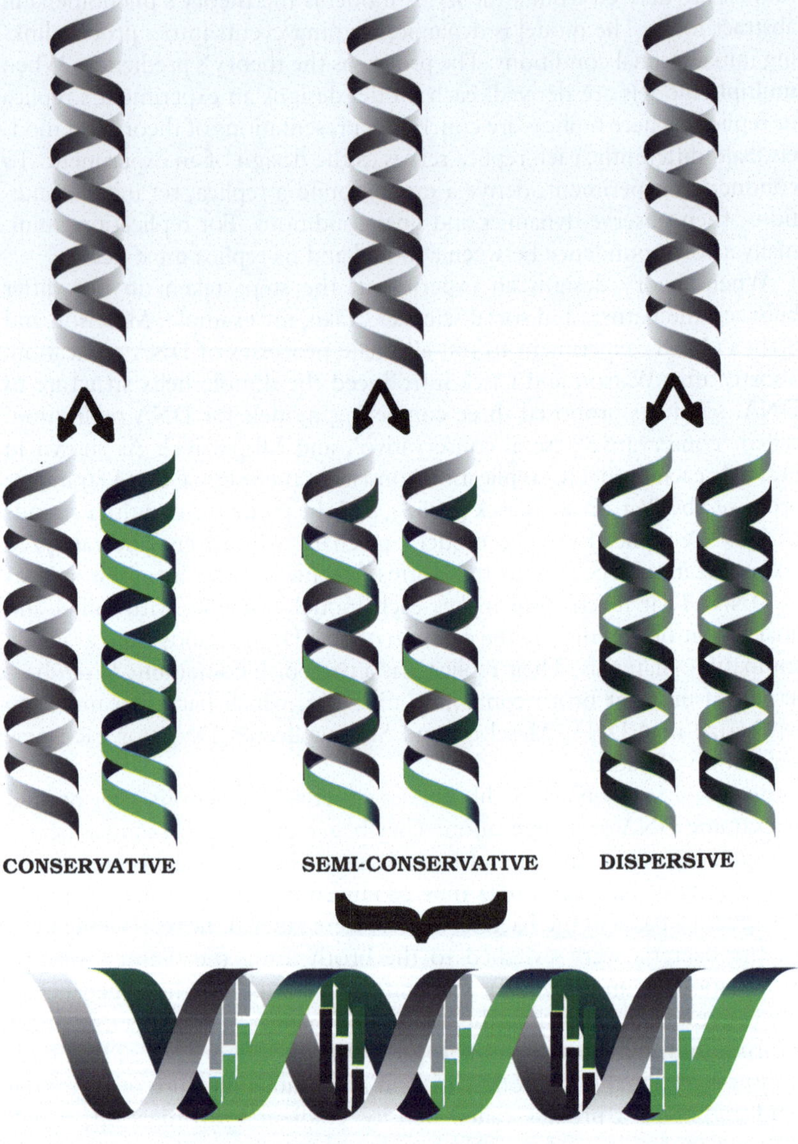

Fig. 4.1 Models of DNA replication

gradient of DNA, to each model's prediction, and the results supported the semi-conservative model's prediction. Here and elsewhere, the results of theory-designed experiments are compared to model predictions, even when multiple replicas and conditions exist.

Turning to social experiments, some basics of Elementary Theory (ET) are now introduced. In ET, social acts, represented as negative and positive sanctions, combine to form events such as exchange in modeled social structures (See Fig. 4.2). Actors take on values and beliefs of the modeled structure, then interact to produce dynamics and outcomes. In models, power differences occur when some actors control and benefit at the expense of others. ET's model of social power draws heavily on Weber's (1918/1968) idea of domination (control) and Marx's (1867/1967) idea of exploitation (power as benefit at another's expense).

In Fig. 4.2, all officials, the Ls, can exchange with M (manager), but Ls in the upper tier set an upper limit on the benefit from exchange of the Ls in the lower tier. Actors in L positions freely move between higher and lower paid positions. In this model, the first two Ls to exchange with M form the upper bound to payoffs to the remaining Ls. The process continues until M's best exchange outcomes for both tiers results (Willer, 1987).

The Fig. 4.2's diagram pictures the elements that make up the exchange structure. For the mechanism, actors' strategically rational decisions animate the negotiation process. The replica includes the model's structure but not its internal dynamics. Subjects make the model dynamic as they decide and act within the structure's positions. Results support theory if the replica's dynamics correspond to those of the model. Further details of this design are given in Chap. 7.

Similar logic guiding the design of physical and social experiments becomes clear when placed side by side. Experiments for both are alike because theory designs both. Investigators derive abstract models from theory, then build corresponding empirical replicas. The mechanism for DNA replication (Watson & Crick, 1953) is enzyme-catalyzed polymerization, and the mechanism for ET's social structure is the rationality-driven acts of subjects. For both, theoretical models are blueprints investigators use to build experimental replicas.

Theories create a well-ordered understanding of the world's phenomena. When a theory succeeds, it makes the world easy to understand because it *imposes* the theory's view on the world. It does so within the process of theory-driven research. In the theory, mechanisms animate abstract models. Replicas bridge the models into the empirical world. The

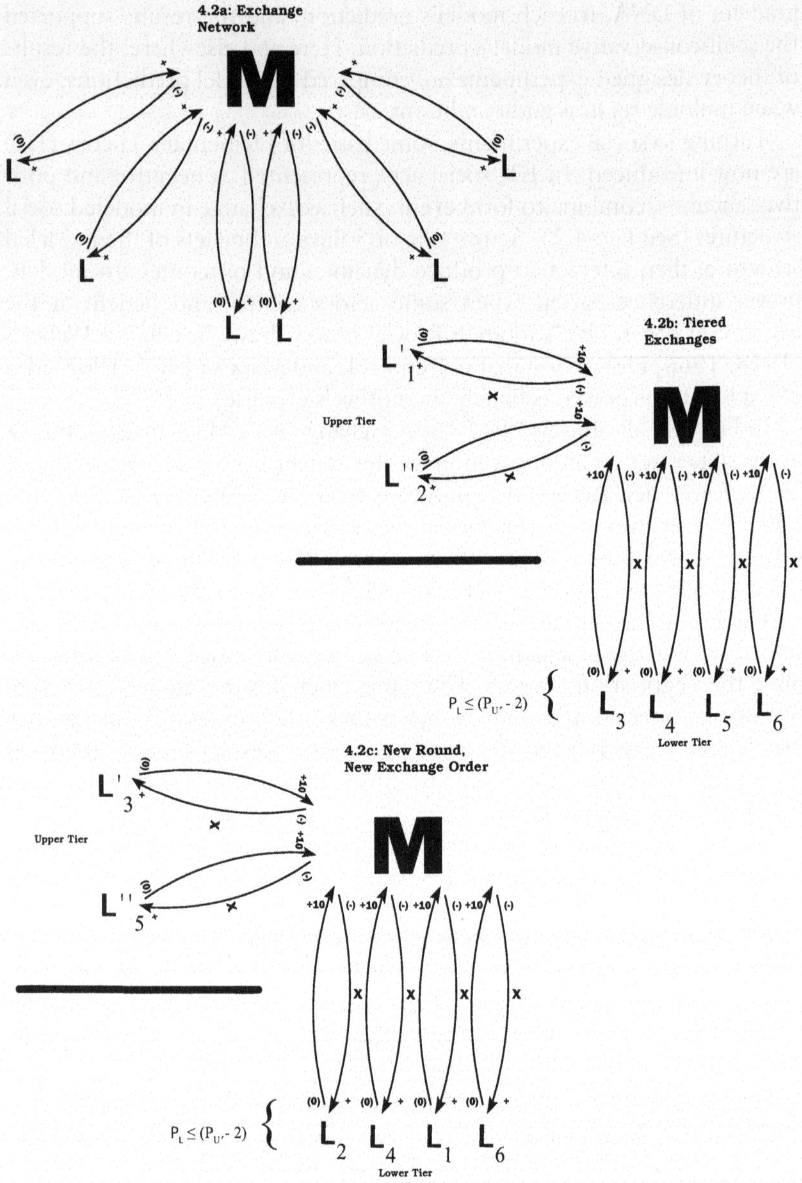

Fig. 4.2 An Experiment on a Tiered Exchange Structure with Mobility

replicas are then set in motion by mechanisms like polymerization or by the decisions and actions of subjects. When outcomes and dynamics of the replica correspond to those of the model, order has been impressed on the world.

Hybrid Experiments and Scientific Exploration

Hybrid experiments combine elements from Fisher-design and theory-designed experiments. The combination can take any number of forms. For example, by violating one or more of a model's assumptions, a hybrid experiment could point out new applications for a theory. Alternatively, hybrids may uncover conditions that limit a model's ability to explain. Here is an outline for a possible hybrid design:

1. Derive a decision model from theory.
2. Derive a second model by altering one condition such as imposing a time limit for decision making.
3. The unaltered model where rational decision making is possible serves as the control condition.
4. Build the two corresponding replicas and run both.

Rand et al. (2012), one of the experiments included in Camerer et al.'s (2016) study, modeled a public goods game. The control condition gave subjects sufficient time to rationally decide whether to contribute and, if so, how much. The experimental condition restricted decision time. Rand et al. (2012) expected that decreasing the time to decide would reduce subjects' rationality and increase their use of intuition (Rand et al., 2012, p. 427). By shortening the subjects' time to decide, they modified a basic assumption of the game-theoretic model. When allowed less decision time, the subjects were significantly more generous. Results suggest (1) that game theory applications should be limited to circumstances where actors can act rationally and (2) that new models for 'intuitive' decision-making need to be created.

Here is a hybrid design that predated the discovery of DNA. Griffith (1928) discovered a bacterial transformation process while researching a pneumonia vaccine. Shown in Fig. 4.3, models of virulent and non-virulent strains of *Streptococcus pneumoniae* differ in the presence of a 'protective polysaccharide coating', which theory argues protects virulent strains from an immune system's attack. Griffith hypothesized that the

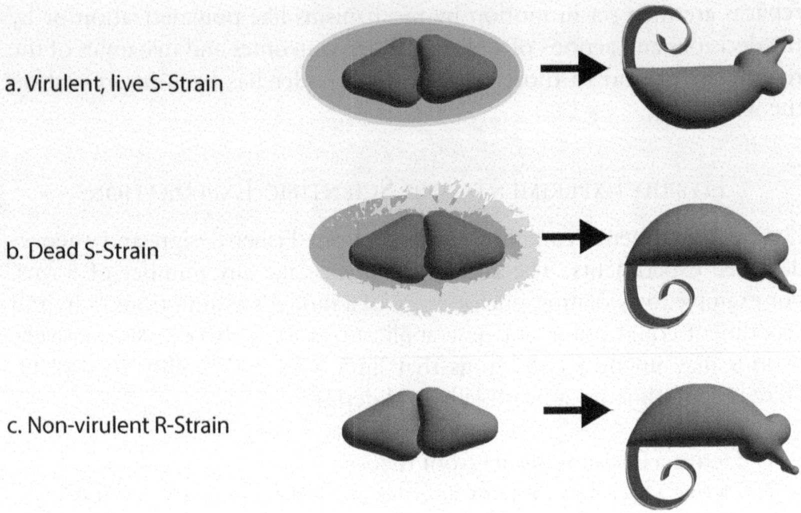

Fig. 4.3 Models of Virulent S-Strain and non-Virulent R-Strain *Streptococcus pneumoniae*

non-virulent R strain could take on properties of the virulent S-strain: he used five conditions to test his hypothesis. Each of his three control conditions conformed to one of the models shown in Fig. 4.3. He injected mice in the three control conditions with either:

1. The non-virulent, live R-strain
2. The virulent, live S-strain
3. The dead S-strains, respectively

The measured end condition was life or death. Theory predicted survival in controls 1 and 3 and death in control 2. The experimental condition (Condition 4) combined the live R-strain and the dead S-strain, and that condition's mice died. Comparing the experimental conditions' results to the control replicas' shows that the live R-strain must have taken on the properties of the virulent S-Strain. The fifth condition provided additional support by extracting the bacteria from the fourth condition's mice and injecting it into new mice. The new mice also died. Though Griffith compared conditions, as in Fisher-design, models of bacterial infection helped interpret results. Notably, the experiment pointed to a 'transforming' mechanism in need of a new theory to explain it.

Griffith's study began a process of scientific discovery. Experiments by Avery et al. (1944) and by Hershey and Chase (1952) built on Griffith's work to identify DNA as the molecule that transfers genetic information. In particular, Avery et al. (1944) used a hybrid experiment to test competing hypotheses, thereby narrowing the possible mechanisms for the transference of genetic information to two. Then, Hershey and Chase (1952) using a theory-driven experiment found support for transference through DNA. The series of studies that traced from Griffith thru Avery et al. and Hershey ending at Watson and Crick illustrates how 'scientific discovery' is a synergistic process wherein theory builds experiments and experiments test and suggest modifications to theory. Scientific knowledge is cumulative. That process is not Mill and Fisher's search for regularities.

More complex hybrid designs change models and replicas in more than one way. For example, Janssen et al. (2010) changed a common pool resource game by introducing costly punishment, an opportunity to communicate with the experiment's other subjects, or a combination of both. The study used a series of six experiments to test each factor and combination of factors. Control and experimental conditions used the same subjects for each experiment in the series. Five subjects collected tokens from a common resource with a density-dependent renewal rate for the control condition. Each collection period lasted four minutes, and the tokens gathered were worth two cents each. Subjects played three consecutive control periods. The control condition was modified by allowing subjects to communicate before collecting tokens, allowing subjects to punish at a cost to self during collection, or both. Results show that communication slows subjects' collecting of tokens, allowing more regenerations, as does punishment. However, when the experimental periods preceded the control periods, cooperation persisted when subjects could no longer communicate but degraded when they could no longer punish.

Here and elsewhere, the defining characteristic of a hybrid experiment is the use of a theory-built replica in the design of the experiment and the differentiation of experimental and control groups by altering one initial condition. Contrasting results for experimental and control groups support (or not) a model's predictions.

Replication Success by Type of Experiment

The authors of this book (Willer & Emanuelson, 2021) reviewed Camerer et al.'s (2016) study to assess replication success by type of experiment. To determine the study type, they applied the following two criteria:

1. The research used one or more models and the corresponding replicas.
2. The research used control and experimental conditions.

Whereas meeting *only* the first criteria indicates a pure theory-design experiment, meeting *only* the second indicates a Fisher-design. If *both* are met, the design is hybrid. Applying the criteria, they discovered that the experiments of Camerer et al. were either Fisher-type or hybrid. There were no theory-driven experiments.

Table 4.1 displays their results. With 6 of the 21 experiments studied being hybrid, and 83% (5 of 6) being successfully replicated. Rand et al. (2012), the only hybrid study not to replicate for Camerer et al., successfully replicated later when Rand (2018) reported results from a new study. By contrast, of the 15 Fisher-type studies, less than 50% replicated. Using

Table 4.1 Frequency of replication by the use of model and replica for designing the experiment

Model & replica	*Replicated*	*Replicated*	*% replicated*
Yes	Balafoutas and Sutter (2012) Derex et al. (2013) Hauser et al. (2014) Janssen et al. (2010) Nishi et al. (2015)	Rand et al. (2012)	83
No	Ackerman et al. (2010) Aviezer et al. (2012) Gneezy et al. (2014) Karpicke and Blunt (2011) Kovács et al. (2010) Morewedge et al. (2010) Wilson et al. (2014)	Duncan et al. (2012) Gervais and Norenzayan (2012) Kidd and Castano (2013) Lee and Schwartz (2010) Pyc and Rawson (2010) Ramirez and Beilock (2011) Shah et al. (2012) Sparrow et al. (2011)	47

a theory-derived model and replica improved the rate of successful replication. Importantly, all hybrid studies tested by Camerer et al. were grounded in game theory.

Experimentation in Economics

Economics experiments were developed in a context very different from that of other social sciences. As Vernon Smith (1989) pointed out, economics is "more theory-intensive and less observation-intensive than perhaps any other science" (p. 151). Consequently, early experimental economists struggled to justify testing theories that other economists considered true *a priori*.

Friedman and Sunder's (1994) *Experimental Methods: A Primer for Economists* indicates the central place of theory in experimental economics. For example, in their Chapter 1, "Theory organizes our knowledge ... theory tells us what data are worth gathering...As theory progresses it guides us...in selecting questions we should ask" (p. 2). In Chap. 2, the authors struggle with the problem of the generality of experimental results. After briefly considering the "problem of induction", they settle for Smith's "parallelism precept" (Friedman & Sunder, 1994, p. 16), that lab-tested propositions also apply outside the lab. In Chap. 3, they take up control issues, both direct and by randomization as well as factorial designs. Their factors are in fact theoretical models. Friedman and Sunder's (1994) concluding chapter compares experimental economics to experimental psychology, finding that theory has a more prominent role in experimental economics (p. 133).

In contrast, in Renita Coleman's *Designing Experiments for the Social Sciences*, theory is not mentioned until Chap. 3, and there, it is (wrongly) defined as "a generalized explanation for some particular phenomena" (2019, p. 57). According to Coleman, experiments provide "insights about the relationship between things where changes in one thing cause something to happen to another". Her Chapter 1 discusses Fisher-type experiment design elements, including variation, confounds, control groups and random assignment. Similarly, Campbell and Stanley's text, *Experimental and Quasi-Experimental Designs for Research,* focuses on design variations of the Fisher-type experiment.

Falsification as a Process

Why do theory-designed experiments replicate, and Fisher-designs do not? Since there is no inductive method, Fisher-design experiments do not produce general knowledge. Therefore, there is no reason why they should replicate. And they usually do not. By contrast, when theory-designed experiments come to replicate, it is because failure to replicate brings about a refinement process that continues until the replication is attained.

Whereas Popper said very little about how theories are discovered—which is to say invented—the authors now build on Popper's idea of falsification by applying it to the process of theory refinement and development. Falsification is the process of refinement wherein theories are applied, falsified, revised, applied again and revised again until further application does not result in falsification. In this process applications may be experiments that have actually been carried out or mental experiments generated in the mind of the theorist or both. In either case, falsifications drive the process forward to the point at which applications do not falsify. Instead, they replicate.

The key to understanding why theories attain replication lies in understanding what happens as the theory is being refined. In fact, two things are happening. First, as the theory is refined, it becomes increasingly insulated from falsification because it is becoming more analytic—which is to say true by definition. That well-formed theories have analytic components is easily seen. For example, force equals mass times acceleration (F = MA) is both a predictive tool and analytically true because force *is* mass times acceleration.

Second, as theory is refined, it becomes more able to impose order on the observations that result from its application. One of the great mysteries of theory construction is which of these two processes is happening in any particular application. When theory begins to replicate, there can be confidence that either it is becoming analytic or becoming more effective at imposing order or a combination of the two. It matters not which of the two processes is occurring. It is enough to know that the theory is becoming more difficult to falsify.

The theory-experiment relation and replication. Figure 4.4 shows that the theory-experiment relation forms a positive feedback loop. Each new cycle produces theory the model of which increasingly resists falsification. The loop continues until successful replication results. Here are the cycle's steps:

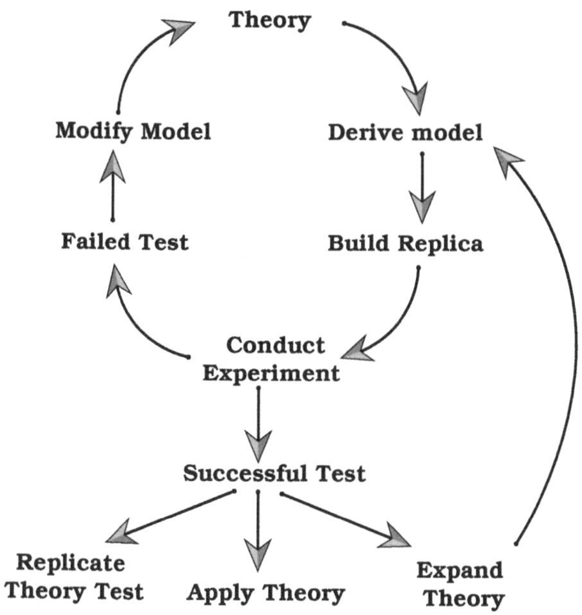

Fig. 4.4 The theory-experiment feedback loop

- Build a model.
- Build a replica.
- Test the replica.
  - If the replica's test fails, revise the model.
  - If the replica's test succeeds,

    proceed to new applications, and
    shift to creating new models within the theory.

As the process continues, knowledge cumulates.

Any number of theories could illustrate the evolving theory-experiment relation. Inheritance theory is tempting. Its progression from Mendel, thru Avery et al. and Hershey, to Watson and Crick is well-documented. Game theory is also possible, as evidence supports its replication success (Willer & Emanuelson, 2021). The authors settled on Elementary Theory (ET) for practical reasons. The first author has personal insight into how it evolved over its four-plus decades.

The authors begin with Elementary Theory's diagraming and modeling procedures, for, as will be seen in Chap. 6, a theory is complete only if it has a procedure to model its phenomena. Physical theories, from Archimedes' laws of levers to Bohr's model of the atom, have a way of picturing their phenomena. Examples in Chap. 6 show further that modeling procedures are commonplace in successful theories. In fact, models are often integral to scientific progress. Watson and Crick (Watson, 1969, p. 130ff) made little progress until they built a 3-dimensional model for the double helix.

ET's modeling procedure borrows much from graph and network theory. The Fig. 4.5 model was the first tested by ET. In the Figure, components such as signs, arcs and actors are combined to diagram exchange relations, which, in turn, combine to create a centralized exchange structure. ET added a mechanism to complete its first model: rational actors acting from positions in the structure. Beliefs and values within the structure motivated actors' actions. In Fig. 4.5, the arcs connecting the A and Bs are called sanctions. Sanctions have + and – signs that indicate value gained or lost when the sanction flows. "Resistance equations", that will be discussed in Chap. 6, link strategic actors' decisions about flows to predict interaction outcomes.

With a model in hand, the next step is to build its replica. Figure 4.5 displays the modeled structure and the room layout, the replica in which

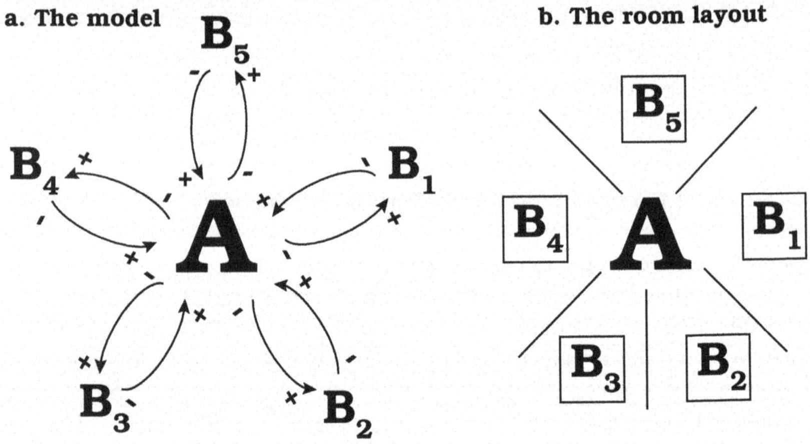

Fig. 4.5 An exchange structure

subjects interacted. Each subject received tokens, and a token's value to each subject matched the values in the structure's model. In the experiment, subjects sent and received offers and completed exchanges. At issue was whether the central position received favorable exchange payoffs.

ET's first test was on the Fig. 4.5's structure. Influenced by the common belief that a position's central location in a network produces a power advantage, it was expected that A would benefit at the expense of each of the peripheral Bs. In fact, central placement did not produce increased benefit. Why not? Perhaps college student participants in the experiment simply did not know how to exercise power. An undergraduate assistant was inserted into the central position in the structure and told to hold out for favorable exchanges. Favorable exchanges were not forthcoming. Instead, participants in peripheral positions simply refused to accept lessor payoffs.

The structure needed to be changed. The change was influenced by Marx's explanation of wages under capitalism. The worker's disadvantage was determined by "the varying proportions in which the working class is divided into active and reserve army" (1867/1967, p. 596). Translated to the lab structure, the central position was limited to fewer exchanges than the number of peripherals available to exchange. That power condition came to be called 'exclusion' a type of connection at the central position. With exclusion, exchange ratios and thus payoffs came to favor the central position.

When a field is scientific, progress in theory comes when favorable results become widely known. Currently sociology is only imperfectly scientific. Falsified formulations frequently linger. In discussing social structures, Walker (2014, p. 128) attributed power to central placement. In fact, it has been known for more than 20 years that central placement is not a power condition. That power was due to exclusion received no mention.

Replication across a theory's scope. As mentioned earlier, replication for theory-designed experiments begins when one model is supported and proceeds through a series of models across the theory's scope. ET developed broader scope by building and testing a series of models. One example is the hierarchy-mobility structure of Fig. 4.2. Weber's discussion on the structure of bureaucracy inspired new models. Of Weber's criteria of bureaucracy, the following were of interest.

1. The organization of offices follows the principle of hierarchy: that is, each lower office is under the control and supervision of a higher one (p. 218).
2. (T)he administrative staff should be completely separated from ownership of the means of administration (pp. 218–219).
3. Candidates are selected on the basis of technical qualifications (p. 220).
4. The salary scale is graded according to rank in the hierarchy (p. 220).
5. There is a system of 'promotion' according to seniority or to achievement, or both. Promotion is dependent on the judgement of superiors (p. 220).

For the model, the manager, M, was given 40 resource units worth one point to M and Ls. Each L was given a resource worth 10 to M and 0 to Ls. The Ls' payoffs were tiered such that the second exchange could earn value equal to or less than the first L to exchange, and payoffs to Ls in exchanges 3–6 could be no greater than two points less than the first L's payoff. Through negotiations with Ls, M decided which L was to exchange first and which exchanged second. L's power moved to the extreme favoring M as Ls competed to move into one of the two upper-tier exchanges.

Subsequent research, discussed in Chap. 7, resulted in a number of further kinds of structural connection with contrasting power effects. These investigations were replications across the theory's scope. ET became able to impose order on the observations across broad scope of structures. Or, perhaps it was simply becoming more analytic. By contrast, without the ability to induce, there can be no corresponding scope-building process for Fisher-design studies. Consequently, when a Fisher-design study fails, there is no falsification process of successive improvement. Instead, research has reached a dead end. Its results go into the file drawer to be forgotten.

There is another kind of scope building. As explained in a previous chapter, Merton (1968/1984) saw "the development of more comprehensive theory as coming about through consolidation of middle range theories" (p. 53). He did not specify a means for consolidating. But modern 'theory-bridges' have joined disparate theories. For example, Elementary theory now integrates parts of Status Characteristics Theory (Simpson & Willer, 2002; Thye et al., 2006; Willer et al., 1997), Legitimacy Theory (Walker & Willer, 2014) and Game Theory (Borch & Willer, 2006). Chapter 9 gives examples of historical explanation that employ multiple theories.

Conclusion

There are two kinds of experimentation. Hence, there are two kinds of replication. The Fisher-design experiment, often seen in social science as the only type of experimentation, uses no theory and is intended to produce generalizations. Not being general, it can only replicate by repeating an original study design as closely as possible. Without a method of induction, however, its generalizations are not general and there is no reason to expect replication to succeed.

Hybrid studies meld elements of Fisher-design and theory-design. They take several forms, but all use replicas to create control and experimental conditions. Hybrid studies contribute to theory development as part of the theory-experiment feedback loop. Importantly, many of the experiments reported by Fisher in his *Design of Experiments* are hybrids.

When theory *designs* experiments, the investigator derives one or more models, then creates a replica for each. A replication is successful if the replica's dynamics and outcome match the dynamics and outcome of the model. Failure to replicate generates a falsification process. When falsification is no longer possible replication succeeds. Then each new application is a further test of the theory as is each new application for explanation or prediction. As a theory develops, its models become resistant to falsification, its scope increases, and knowledge cumulates.

Because there is no inductive method, no amount of refinement can make Fisher-designs replicate such that the knowledge produced is useful. Explanation and prediction cannot use generalizations, the replication rates of which are, at best, hardly better than a coin flip. The inability to consistently replicate is a problem across social science only to the extent that the Fisher-design is the dominant method. Fisher-design experiments are not the only kind of experiment, but they are the only kind that does not work.

References

Ackerman, J. M., Nocera, C., & Bragh, J. A. (2010). Incidental haptic sensations influence social judgments and decisions. *Science, 328*(5986), 1712–1715.

Anvari, F., & Lakens, D. (2018). The replicability crisis and public trust in psychological science. *Comprehensive Results in Social Psychology, 3*(3), 266–286. https://doi.org/10.1080/23743603.2019.1684822

Avery, O. T., McLeod, C. M., & McCarty, M. (1944). Studies on the chemical nature of the substance inducing transformation of pneumococcal types. *The Journal of Experimental Medicine, 79*(2), 137–158. https://doi.org/10.1084/jem.79.2.137

Aviezer, H., Trope, Y., & Todorov, A. (2012). Body cues, not facial expressions discriminate between positive and negative emotions. *Science, 338*(6111), 1225–1229.

Baker, M. (2016). 1,500 scientists lift the lid on reproducibility. *Nature, 533*, 452–454. https://doi.org/10.1038/533452a

Balafoutas, L., & Sutter, M. (2012). Affirmative action policies promote women and do not harm efficiency in the laboratory. *Science, 335*(6068), 579–582.

Benjamin, D. J., Berger, J., Johannesson, M., Nosek, B. A., Wagenmakers, E., Berk, R., ... Johnson, V. (2017, July 22). Redefine statistical significance. https://doi.org/10.31234/osf.io/mky9j

Borch, C., & Willer, D. (2006). Power, embedded games, and coalition formation. *Journal of Mathematical Sociology, 30*, 77–111.

Camerer, C. F., Dreber, A., Forsell, E., Ho, T. H., Huber, J., Johannesson, M., et al. (2016). Evaluating replicability of laboratory experiments in economics. *Science, 351*(6280), 1433–1436.

Casey, K., Glennerster, R., & Miguel, E. (2012). Reshaping institutions: Evidence on aid impacts using a preanalysis plan. *The Quarterly Journal of Economics, 127*(4), 1755–1812.

Coleman, R. (2019). *Designing experiments for the social sciences.* Sage.

Derex, M., Beugin, M.-P., Godelle, B., & Raymond, M. (2013). Experimental evidence for the influence of group size on cultural complexity. *Nature, 503*(7476), 485–487.

Duncan, K., Sadanand, A., & Davachi, L. (2012). Memory's penumbra: Episodic decisions induce lingering mnemonic biases. *Science, 337*(6093), 485–487.

Earl, B. D., & Trifimow, D. (2015). Replication, falsification, and the crisis of confidence in social psychology. *Frontiers in Psychology, 6*, 621.

Ferguson, C. J., & Heene, M. (2012). A vast graveyard of undead theories: Publication bias and psychological science's aversion to the null. *Perspectives on Psychological Science, 7*(6), 555–561. https://doi.org/10.1177/1745691612459059

Franco, A., Malhotra, N., & Simonovits, G. (2014). Publication bias in the social sciences: Unlocking the file drawer. *Science, 345*(6203), 1502–1505.

Franklin, A. (2018). *Is it the 'same' result: Replication in physics.* Iop Concise Physics.

Friedman, D., & Sunder, S. (1994). *Experimental methods: A primer for economists.* Cambridge University Press.

Galileo, G. (1954). *Dialogues concerning two new sciences* (H. Crew & A. de Salvio, Trans.). Dover. (Original work published ca. 1638).

Gerber, A. S., & Malhotra, N. (2008). Publication bias in empirical sociological research: Do arbitrary significance levels distort published results? *Sociological Methods & Research, 37*(1), 3–30. https://doi.org/10.1177/0049124108318973

Gervais, W., & Norenzayan, A. (2012). Analytic thinking promotes religious disbelief. *Science, 336*(6080), 493–496.

Gneezy, U., Keenan, E. A., & Gneezy, A. (2014). Avoiding overhead aversion in charity. *Science, 346*(6209), 632–635.

Griffith, F. (1928). The Significance of Pneumococcal Types. *The Journal of Hygiene, 27*(2), 113–159. https://doi.org/10.1017/s0022172400031879

Hauser, O., Rand, D., Peysakhovich, A., & Nowak, M. (2014). Cooperating with the future. *Nature, 511*(7508), 220–223.

Hershey, A. D., & Chase, M. (1952). Independent functions of viral protein and nucleic acid in growth of bacteriophage. *The Journal of General Physiology, 36*, 39–56.

Janssen, M. A., Holahan, R., Lee, A., & Ostrom, E. (2010). Lab experiments for the study of social-economic systems. *Science, 328*(5978), 613–617.

Karpicke, J. D., & Blunt, J. R. (2011). Retrieval practice produces more learning than elaborate studying with concept mapping. *Science, 331*(6018), 772–775.

Kidd, E., & Castano, D. (2013). Reading literary fiction improves theory of mind. *Science, 342*(6156), 377–380.

Killen, P. R. (2005). An alternative to null-hypothesis significance tests. *Psychological Sciences, 16*(5), 345–353. https://doi.org/10.1111/j.0956-7976.2005.01538.x

Klein, R. A., Vianello, M., Hasselman, F., Adams, B. G., Adams, R. B., et al. (2018). Many labs 2: Investigating variation in replicability across samples and settings. *Advances in Methods and Practices in Psychological Science, 1*(4), 443–490. https://doi.org/10.1177/2515245918810225

Kovács, Á. M., Téglás, E., & Endress, A. D. (2010). The social sense: Susceptibility to others' beliefs in human infants and adults. *Science, 330*(6012), 1830–1834. https://doi.org/10.1126/science.1190792. PMID: 21205671.

Lee, S. W. S., & Schwartz, N. (2010). Washing away post decisional dissonance. *Science, 328*(5979), 709.

Lieberson, S., & Lynn, F. B. (2002). Barking up the wrong branch: Scientific alternatives to the current model of sociological science. *Annual Review Sociology, 28*, 1–19.

Marx, K. (1967). *Capital*. International Publishers. (Original work published 1867).

Merton, R. K. (1984). *Social theory and social structure*. The Free Press. (Original work published 1968).

Meselson, M. S., & Stahl, F. W. (1958). The replication of DNA in *Escherichia coli*. *Proceeding of the National Academy of Sciences USA, 44*(7), 671–682.

Moody, J. W., Keister, L. A., & Ramos, M. C. (2022). Reproducibility in the social sciences. *Annual Review of Sociology, 48*, 65–85.
Morewedge, C. K., Eun, Y. H., & Vosgerau, J. (2010). Thought for food: Imagined consumption reduces actual consumption. *Science, 330*(6010), 1530–1533.
Muthukrishna, M., & Henrich, J. (2019). A problem in theory. *Nature: Human Behavior, 3*, 221–229.
Newton, I. (1966a). *Mathematical principles of natural philosophy and his system of the world: The motion of bodies* (A. Motte, Trans. 1729, Vol. 1). University of California Press. (Original work published 1686).
Newton, I. (1966b). *Mathematical principles of natural philosophy and his system of the world: The system of the world* (A. Motte, Trans. 1729, Vol. 2). University of California Press. (Original work published 1687).
Nishi, A., Shirado, H., Rand, D., & Christakis, N. (2015). Inequality and visibility of wealth in experimental social networks. *Nature, 526*(7573), 426–431.
Open Science Collaboration. (2015). Estimating the reproducibility of psychological science. *Science, 349*(6251), 1–8.
Popper, K. R. (1957). *The poverty of historicism*. Harper & Row Torchbooks.
Pyc, M. A., & Rawson, K. A. (2010). Why testing improves memory: Mediatory effectiveness hypothesis. *Science, 330*(6002), 335.
Ramirez, G., & Beilock, S. L. (2011). Writing about testing worries boosts exam performance in the classroom. *Science, 331*(6014), 211–213.
Rand, D. G. (2018). Non-naïvety may reduce the effect of intuition manipulations. *Nature: Human Behavior, 2*(9), 602.
Rand, D. G., Greene, J., & Nowak, M. (2012). Spontaneous giving and calculated greed. *Nature, 489*(7416), 427–429.
Ravetz, J. R. (1971). *Scientific knowledge and its social problems*. Oxford University Press.
Resnick, B. (2018, April 27). More social science studies just failed to replicate. Here's why this is good. *Vox*. https://www.vox.com/science-and-health/2018/8/27/17761466/psychology-replication-crisis-nature-social-science
Rosenthal, R. (1979). The file drawer problem and tolerance for null results. *Psychological Bulletin, 86*(3), 638–641.
Shah, A. K., Mullainathan, S., & Shafir, E. (2012). Some consequences of having too little. *Science, 338*(6107), 682–685.
Silberzahn, R., & Uhlmann, E. (2015). Crowdsourced research: Many hands make tight work. *Nature, 526*, 189–191. https://doi.org/10.1038/526189a
Simpson, B., & Willer, D. (2002). Applications of elementary theory to the social structures of antiquity. In S. C. Chew & J. D. Knottnerus (Eds.), *Structure, culture and history: Recent issues in social theory* (pp. 231–254). Rowman and Littlefield.

Smith, V. (1989). Theory, experiment, and economics. *Journal of Economic Perspectives, 3*(1), 151–169.

Sparrow, B., Liu, J., & Wegner, D. M. (2011). Google effects on memory: Cognitive consequences of having information at our fingertips. *Science, 333*(6043), 776–778.

Starbuck, W. H. (2016). 60th anniversary essay: How journals could improve practices in social sciences. *Administrative Science Quarterly, 61*(2), 165–183.

Steinle, F. (2016). Stability and replication of experimental results: A historical perspective. In H. Atmanspacher & S. Maasen (Eds.), *Reproducibility: Principles, problems, practices and prospects* (pp. 39–64). John Wiley.

Thye, S. R., Willer, D., & Markovsky, B. (2006). From status to power: New models at the intersection of two theories. *Social Forces, 84*, 1471–1496.

Turner, J. H. (1993). *Classical sociological theory: A positivist's perspective.* Nelson-Hall, Inc.

Walker, H. A., & Willer, D. (2006). Experimental methods. In G. Ritzer (Ed.), *Encyclopedia of sociology* (Social Psychology). Blackwell Publishing.

Walker, H. A., & Willer, D. (2007). Experiments and the science of sociology. In M. Webster & J. Sell (Eds.), *Laboratory experiments in the social sciences* (pp. 25–54). Elsevier Inc.

Walker, H. A., & Willer, D. (2014). Legitimizing collective action and countervailing power. *Social Forces, 92*, 1217–1239. https://doi.org/10.1093/Sf/Sot116

Walker, L. S. (2014). Developing your experiment. In M. Webster Jr. & J. Sell (Eds.), *Laboratory experiments in the social sciences* (2nd ed., pp. 127–144). Elsevier.

Watson, J., & Crick, F. (1953). Genetical implications of the structure of deoxyribonucleic acid. *Nature, 171*, 964–967. https://doi.org/10.1038/171964b0

Watson, J. D. (1969). *The double helix: A personal account of the discovery of the structure of DNA.* Signet.

Weber, M. (1968). *Economy and society.* University of California Press. (Original work published 1918).

Willer, D. (1987). *Theory and the experimental investigation of social structures.* Gordon and Breach Science Publishers.

Willer, D., & Emanuelson, P. (2021). Theory and the replication problem. *Sociological Methodology, 51*(1), 146–165.

Willer, D., Lovaglia, M., & Markovsky, B. (1997). Power and influence: A theoretical bridge. *Social Forces, 76*, 571–604.

Willer, D., & Walker, H. A. (2007). *Building Experiments.* Stanford University Press.

Wilson, T. D., Reinhard, D. A., Westgate, E. C., Gilbert, D. T., Ellerbeck, N., Hahn, C., Brown, C. L., & Shaked, A. (2014). Social psychology. Just think: The challenges of the disengaged mind. *Science, 345*(6192), 75–77. https://doi.org/10.1126/science.1250830

PART II

Theory-Driven Research in Social Science

CHAPTER 5

Worldviews

The paramount goal of all sciences is the discovery of general knowledge. With general knowledge comes opportunities to explain, predict and, for historical events, postdict. As Einstein pointed out, theory is the free invention of the human mind. The *discovery* of general knowledge is through the *invention* of theory. In fact, discovery of theory and invention of theory are the same process. Given the importance of theory, it is surprising that almost nothing has been written about the components of theory and how they are deployed. As seen in Chap. 3, leading philosophers of science said surprisingly little about theory, its components and how it works. They are not alone. No one else has written in detail about how theory works.

With so little previously said about how to do theory, it is necessary to begin from ground zero. Ground zero was exactly the point of departure from which the authors started building and applying a social theory. Here are the components of theory that were discovered.

The Four Components of Scientific Theory

1. Worldviews that conceptualized phenomena
2. Modeling procedures that picture the phenomena
3. Mechanisms and laws that render the picture dynamic
4. Experimental replicas as designed by theory

D. Willer, P. Emanuelson, *The Logic of Theoretical Sociology*,
https://doi.org/10.1007/978-3-031-91709-7_5

All fully developed scientific theories have the same four components. This chapter focuses on the first of those components, the worldview (in German, *weltanschauung*). Worldviews conceptualize phenomena for explanation and prediction. In doing so, they define the phenomenon that the theory is to explain and indicate how to frame that explanation. Sometimes they include mechanisms that allow predictions. Exactly how worldviews work is shown here by contrasting the worldview of Aristotle with that of Galileo and Newton. Beginning with physical theory is justified because all fully developed scientific theories have the same basic structure. Furthermore, the contrasts between Aristotle on the one hand, and Galileo and Newton on the other drive home the nature and significance of worldviews for theory. Having begun with examples from classical physics, the discussion moves to four worldviews that have been proposed for social theory, those of Marx, Weber, Homans and Coleman.

In examining worldviews, it needs to be emphasized that this is not a general review of the work of any of the scholars whose work is discussed. To the contrary, the aim is only to uncover how each saw the part of the world about which they were theorizing, nothing more.

Aristotle

The cornerstone of Aristotle's physics is that motion of a body is through another body such as a boat moving through water or an arrow through air. "A body is in a place if another body surrounds it; otherwise, it is not" (Aristotle, 330 BCE/1961, p. 66). Since motion always involves one body moving through another (Aristotle, 330 BCE/1961, p. 72), motion continues as long as the body is pushed. "(S)uch a mover does this by contact" (Aristotle, 330 BCE/1961, p. 44). Movement is going from place to place. Since motion stops when pushing stops, projectiles are a problem. For example, a rock continues in motion after it is thrown. Aristotle suggests it may be that when "air which has been pushed pushes projectiles with a motion more vigorous than their motion to their resident place" (Aristotle, 330 BCE/1961, p. 72). The projectile stops when it has reached its resident place. Similarly, a falling body stops when it hits the ground because the ground is its resident place.

When bodies move, that motion is against resistance of the body through which they are moving (Aristotle, 330 BCE/1961, p. 72). Bodies move at different speeds for two reasons:

> either, on the one hand, because of a difference in the medium, for example, in earth or water or air; or, on the other hand, because of a difference in the moving bodies compared...one body is heavier or lighter than the other. (Aristotle, 330 BCE/1961, p. 72)

Furthermore, motion through a denser body is slower than through a less dense one (Aristotle, 330 BCE/1961, p. 73).

For centuries, those who followed Aristotle were notorious for asserting that heavier bodies fell faster than lighter bodies. It was actually Aristotle's relation between weight and speed of falling that Galileo falsified. The authors of this book searched through Aristotle's *Physics*, but did not uncover even one assertion concerning the speed of falling bodies. Nor is the phrase "falling bodies" or the equivalent ever mentioned in Aristotle's *Physics*. What then is the source of the belief that Aristotelian physics holds that bodies fall at a rate proportional to their weight? It may follow from this passage:

> We see that, other things being equal, heavy and light bodies move with unequal velocities over an equal space in the ratio which their magnitudes have to each other. Hence, they must do so even when moving through a void. But this is impossible. Why should one of them move faster than another? To be sure, one of them does necessarily move faster in a space full of matter because the greater body divides the medium faster by reason of its force; for a body moving naturally or a projectile divides a medium as it does by reason either of its shape or of its upward or downward tendency. (Aristotle 330 BCE/1961, p. 74 [brackets removed])

If "a body moving naturally" means a freely falling body, as the authors suspect, then Aristotle is asserting that, all else being equal, bodies *fall* at a rate proportional to their weight.

Since the cornerstone of Aristotle's physics is that motion of a body is always through another body, motion is never through a void. Thus, from the worldview of his theory, motion and resistance to motion by a medium are indivisible. That view is in complete contrast to that of Galileo and Newton for whom natural motion is through the nothingness of geometric space. Furthermore, if motion through a void is possible, Aristotle's physics would fall apart. As a result, it is not surprising that a number of quite long passages in his *Physics* are concerned with "proofs" that a void is impossible.

From the point of view of Aristotle, it is quite reasonable to think of bodies falling at different rates with heavy bodies falling faster than lighter ones. And he knew full well that at least some of that difference was due to resistance of the medium through which the body was falling. Elsewhere in *Physics* he states: "when things are set in motion by something else" it is by pushing, pulling, carrying or rotation (132). Since that motion is produced by contact, it is by no means clear why anything falls. What form of contact is pushing or pulling falling bodies? It must be then that falling is due to "action at a distance" for which there is little or no room in his physics. It is only fair to mention that both Newton and Einstein also abhorred action at a distance and worked hard to eliminate it from their theories.

Late in *Physics* Aristotle suggests that anything in motion is set in motion by another thing. In turn that other thing was set in motion by yet another and so forth. Asserting that infinite regress is impossible, there must have been an unmoved mover. In later centuries, Church fathers used this passage as a part of proof of the existence of God.

At the end of our copy of *Physics* is found a listing of Aristotle's technical terms, 211 in number. What we have learned since Aristotle is to be far more parsimonious with technical terms. As Einstein said, a theory should have as few technical terms as possible (1954, p. 272)—which is good advice to any theorist.

Given Aristotle's worldview, motion in a void is nonsense. Aristotle asserts that, in a void: "Hence a body would either continue in its state of rest or would necessarily continue in its motion indefinitely, unless interfered with by a stronger force" (Aristotle 330 BCE/1961, p. 72). Students of the history of science will immediately recognize that quote from Aristotle as Newton's First Law. (See below.) That it is basic to Newton's theory of motion and nonsense within Aristotle's worldview are both true—and illustrate well the role and significance of worldviews, of *weltanschauungs* in scientific theory.

Galileo Galilei and Isaac Newton

The cornerstone of Galileo's and Newton's physics is the 'theoretically pure' case of a body in motion through empty "geometric" space. Their empty space is the void that Aristotle railed against. Of Galileo's *Two New Sciences* (1638/1954) the first was concerned with the strength of various shapes and the second with the motion of bodies. Let's look only at the second beginning with the speed of falling bodies. It will be remembered that Aristotle held that bodies have natural rates of fall proportionate to their weight. To falsify Aristotle, Galileo (1638/1954, pp. 62–63) suggested that two bodies with different rates fall be attached to each other. With the lighter body's natural rate of falling being slower and the heavier body faster, the composite body should fall at a speed between the two. However, the composite body, being heavier than either of its parts, must fall faster. Therefore, the Aristotelian view is self-contradictory. Next Galileo reported experiments with two iron balls of the same density and found that the larger fell only very slightly faster than the smaller. He then turned to the density (what he called the specific gravity) of the medium through which bodies fall.

> Because if we find as a fact that the variation of speed among bodies of different specific gravities is less and less according as the medium becomes more and more yielding, and if finally in a medium of extreme tenuity, though not a prefect vacuum, we find that, in spite of great diversity of specific gravity, the difference in speed is very small and almost inappreciable, then we are justified in believing it highly probable that in a vacuum all bodies would fall at the same speed. (p. 72)

Of course, Galileo knew that falling bodies did not fall at a uniform speed. They accelerated. By "naturally accelerated" he referred to motion that "when starting from rest, it acquires, during equal time-intervals, equal increments of speed" (p. 162). What proof did he offer? The proof lay, not in rational argument, but in a series of experiments using a specially designed and constructed apparatus, an inclined plane. The apparatus solved the problem that bodies fell too fast for the rate of acceleration to be accurately measured. He slowed the process by substituting balls rolling down an inclined plane for freely falling balls. He then devised a new water clock to measure time of descent.

Having carried out a number of experiments empirically validating uniform acceleration, he turned back to theory to untangle the problem of projectiles. Projectile motion is "compounded of one which is uniform and horizontal and of another which is vertical and naturally accelerated" (p. 244). Figure 5.1 shows Galileo's figure for how the two motions are compounded and how the result is a semi-parabola. In Fig. 5.1 motion is from right to left.

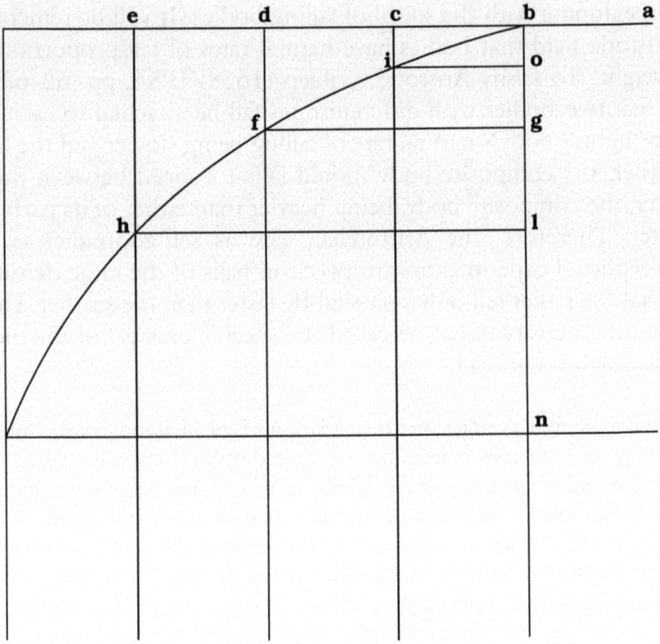

Fig. 5.1 Galileo's diagram of projectile motion

Galileo's thinking here is fully modern and clearly different from ancients such as Aristotle. First, he moved from theoretical inference to experimental proof, then back to theory, and back to experiment. Thus, he built complexity step by step. Second, he established significance for his work by falsifying Aristotle's widely held alternative position. Third, his new idea of compounding different kinds of motion solved the problem of projectiles by converting motion into a problem of geometry. Together these three Galilean innovations were used again by Newton.

Reading Newton after having read Galileo reveals a continuity between the two. Newton carried the Galilean theory of the motion of bodies to new levels of precision and generality. For example, Newton's first law is:

> *Every body continues in its state of rest, or of uniform motion in a right line, unless it is compelled to change that state by forces impressed upon it.* (1686/1966a, p. 13 [italics original])

This law seems to be a straightforward generalization of Galileo's solution for projectiles. Certainly, the diagram Newton draws for the joint effect of two forces (1668/1966b, p. 14.) is an extension of Galileo's compounding (see Fig. 5.2).

But Newton did something that Galileo had not attempted. He integrated the motion of bodies with planetary motion. The core diagram for that integration is given here as Fig. 5.3. That figure is read as follows. A person at point V throws a series of balls horizontally, each at a higher speed. The paths of the A, B, C and D balls are larger and larger parabolas. Finally, a ball is thrown hard enough that it returns to point V. That path and the paths beyond it are the ellipses of planetary motion predicted by Kepler. Thus did Newton integrate local and planetary motions.

Changes in worldviews about the motion of bodies demonstrated two progressive changes. First, as worldviews evolved, the world being pictured became simpler. Certainly, Galileo's motion through nothingness is much simpler than Aristotle's motion of a body within a body. Displacing Aristotle, Galileo's simpler motion became the center of focus while issues such as friction shifted to the periphery, became less important, and were left for a future time to resolve—if deemed as worthy of resolution. Second, beyond being simpler, the new worldview was less 'realistic', less a carbon copy of the world as it first appears to us.

The idea that conceptions could purposefully be less realistic is not unknown to sociology. Weber said it well when referring to ideal types as incorporating an "one-sidedly emphasized viewpoint." Of course, not just any "unrealistic view" will result in a powerful theory. It seems that Einstein worked long and hard to discover the "one-sidedly emphasized viewpoint" he needed for his theory. Being simpler and being less realistic are related processes. Interestingly, when a theory is particularly successful, its worldview can come to be incorporated into the dominant culture

Fig. 5.2 Newton's diagram for the effect of Joint Forces N & M

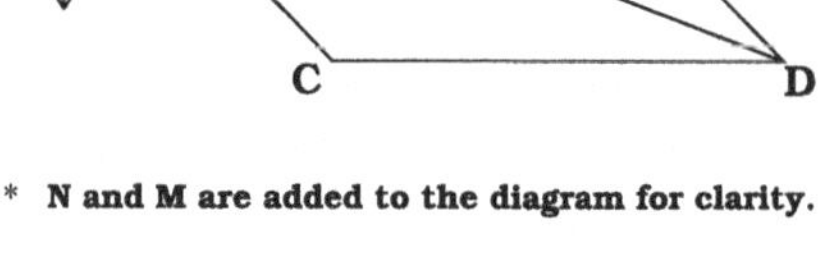

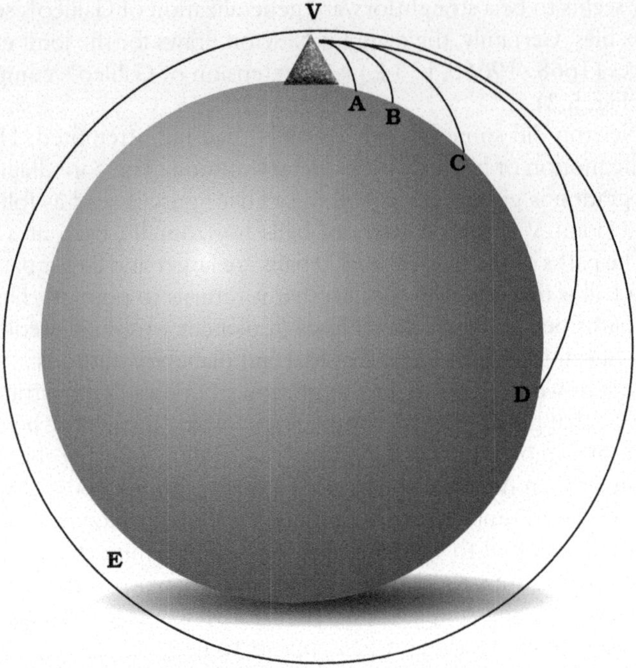

*Modified from Newton's figure of Centripetal Force in Principia Mathematica

Fig. 5.3 Newton's action of centripetal force

with the result that its sense of one-sidedness fades or is lost. For example, today the dominant view of bodies in motion is Newtonian, not Aristotelian.

Karl Marx

Compared to worldviews about bodies in motion, worldviews in sociology have the widest imaginable diversity. Phenomena range from small groups to huge power structures. Social structures at any place on the globe and any time in human history and prehistory are all legitimate subjects for study. Looking at four worldviews proposed for sociology, of particular interest will be whether those views simplify and what kind of one-sidedness is being offered. Worldviews in sociology occasionally confront each other but, when they do, may not displace each other.

Given the scope of his theorizing, it is difficult to characterize Marx's worldview. But fortunately, the man himself can help. In the Preface to *A Contribution to the Critique of Political Economy* Marx offered a summary of his guiding principles, which is to say his worldview. In reading Marx's worldview—and others to follow—it would be good to keep in mind that simplicity is a virtue and realism may not be. Here are the crucial parts of his Preface.

> In the social production of their existence, men inevitably enter into definite relations, which are independent of their will, namely relations of production appropriate to a given stage in the development of their material forces of production. The totality of these relations of production constitutes the economic structure of society, the real foundation, on which arises a legal and political superstructure and to which correspond definite forms of social consciousness...It is not the consciousness of men that determines their existence, but their social existence that determines their consciousness... No social order is ever destroyed before all the productive forces for which it is sufficient have been developed, and new superior relations of production never replace older ones before the material conditions for their existence have matured within the framework of the old society. (1859/2009, pp. 20–21)

Marx's worldview is deterministic. Indeed, determinism is one of the great virtues of his theory. Nor is determinism a fault in any social theory. To the contrary, every social theory must see human activity as determined at some level. If a theory does not, it says nothing whatsoever about how people act and why they act. Saying nothing, they are not theories. Nor is Marx alone in his determinism. The recent history of experimental research in sociology is a history of better and better predictions of human activity—of more and more precise grasp of how human activity is determined. (See Chap. 8.)

Marx begins with humans entering relations that are preset, not freely created by those acting in them. It is within these relations that determinism first occurs. Taken together, these relations are the economic structure, the foundation on which the other parts of the social structure are built. It is between the economic structure and the other elements of society, including consciousness, that the second level of determinism occurs. Having found two levels of determination, at issue is how each of the two works. Let's look at determination in relations first.

Marx's assertion that existence determines consciousness and not the reverse helps to conceptualize human action in social structures. The key is to analyze consciousness into its component parts: values, beliefs and a mechanism for decisions. Marx is clear that values are acquired from the relations in which actors are engaged. It follows that, if the structure is known, values of those in the structure can be inferred. Similarly, beliefs are acquired from the structure. The beliefs of each actor include values and beliefs of other actors. But beliefs can pose a problem for theory. To be complete, each actor's beliefs must contain the beliefs of others about self. However, others' beliefs about self also contain others' beliefs about self's beliefs about others' beliefs—and similarly *ad infinitum*. Marx cut through this regress by having all actors' beliefs determined by a common factor. That common factor is 'conditions of existence'. With values and beliefs being structurally determined, adding general decision procedures gives the theoretic actor with consciousness determined by existence.

The second level of determination occurs between parts of the social structure. Those changes are qualitative. Qualitative change occurs through a transition in which one epoch together with its mode of production falls and a second rises. Transitions occur when new, superior forces of production have grown within the framework of the old order. Marx offers no general answer to exactly how new structures grow within the old order. He has, however, much to say specifically about the transition from feudalism to capitalism.

That transition begins with 'primitive accumulation' so-called "because it forms the pre-historic stage of capital" (Marx 1867/1967, p. 668). "The so-called primitive accumulation, therefore, is nothing else than the historical process of divorcing the producer from the means of production" (p. 668). To which he adds, "The economic structure of capitalistic society has grown out of the economic structure of feudal society. The dissolution of the latter set free the elements of the former" (p. 668).

The rise of the wage laborer and capitalist are linked processes. The key to both is the separation of the worker from the means of production, means that come to be held by the capitalist. Marx traces that separation in England toward the end of Volume 1 of *Capital*. Under Elizabethan labor laws, people were forcibly expropriated from the land to become beggars or vagabonds, which is to say to become proletariat (p. 686ff). "The bourgeoisie, at its rise, wants and uses the power of the state to 'regulate wages'" (p. 689). At the beginning of the transition, wages were not determined by the existence of the 'reserve industrial army' but by the coercive power of the state.

The separation of the worker from the means of production initiates the *structural power* of capitalist class over the working class. That separation creates an exchange relation between workers and capitalists in which, under capitalism, there is a surplus of workers over available jobs. That surplus means that, in the absence of unions, there is no bargaining in the worker—capital relation. Instead, workers offer to work at wages better and better to the capitalist until equilibrium is reached at subsistence (Marx 1867/1967, p. 596). Marx understood the importance of the surplus of workers over jobs and recognized the need for a theory of population that would explain that surplus. As he said, "every special historic mode of production has its own special laws of population" (p. 592). Today, regardless of population change, the state regulates exchange rates of employment with the policy that they are set by some non-zero level of unemployment—by a contemporary reserve industrial army. At least that is how it works in the U.S.

While Marx's theory has certainly been employed for ideological purposes, that use should not blind the theorist to its utility as a worldview or its potential explanatory power. To dismiss Marx as a mere ideologue is the equivalent of dismissing Galileo as a heliocentrist.

Max Weber

The sociologist whose scope of knowledge is comparable to that of Marx is Max Weber. Weber's works extended from ancient economic history, through the religions of China and India, the fall of Rome, political and economic developments in Europe, and the analysis of types of organization. To the authors there is no passage in Weber comparable to Marx's building principles. Nevertheless, the following captures something of Weber's worldview.

> Not ideas, but material and ideal interests, directly govern men's conduct. Yet very frequently the 'world images' that have been created by 'ideas' have, like switchmen, determined the tracks along which action has been pushed by the dynamic of interest. (Weber, 1922/1946, p. 280)

In Weber's steam engine of history, ideas are not a straightforward reflection of existence nor are material interests eclipsed by prevailing ideas. Historical change is jointly affected by material and ideal interests. Furthermore, the direction of effect of economic structures on all else in society is by no means one-way. For example, Weber would see the impact of Elizabethan labor laws on the economy as evidence that the state can produce economic change. By contrast, Marx saw the rising bourgeoisie using the state to separate the workers from the means of production.

How systematic are Weber's diverse investigations? Seeking hints of system, let's turn to how Weber introduced his basic concepts. That introduction is found in the first volume of *Economy and Society* that, as students of Weber know, was written after the second volume. The systematic way to introduce concepts is to begin with the simplest and build to more and more complex formulations. Let us see what Weber did.

Weber's beginning point is social action. Action is social because the actor's 'meaning' takes into account the behavior of others (Weber, 1918/1968, p. 4, 22). Weber's meaning, if not identical to Marx's consciousness, is similar in that, for both, beliefs and values are attributed to actors. We note that the potential of infinite regress in actors' beliefs does not have Marx's solution in Weber. The next step of complexity beyond social action is social relationship. Weber introduces 'social relationship' in the following way.

> The term "social relationship" will be used to denote the behavior of a plurality of actors insofar as, in its meaningful content, the action of each takes account of that of the others and is oriented in these terms. (1918/1968, p. 26)

Note that Weber packed multiple actors into a single social relation. Beginning with action, then introducing relationship, the next step was to introduce structural-level formulations. Here he introduced types of legitimate order and organizations.

At this point it is increasingly obvious that, other than the order in which they are introduced, there is little that is systematic about the way Weber introduces his terms. Weber was right to begin with the simplest formulation, social action, then introduce social relation and finally social structures. It would have been more systematic, however, to build more complex concepts by defining them using only the simpler concepts already introduced.

Building concepts step by step and defining more complex formulations by simpler ones is both more parsimonious, and more precise. For example, in our view, meaning is not a unitary concept. It, like consciousness, has parts: the actor's values, beliefs and capability of deciding. Then a social relationship exists when two actors act meaningfully with each other. Each takes the values and beliefs of the other into account. Social structure is defined as two or more connected social relations. Better yet would be to introduce geometric representations parallel to these definitions. Then one could *look at* social structures just as Galileo and Newton *looked at* representations of motion. Parallel geometric representations have a further advantage. When looking at social structures composed of connected social relations, it is natural to ask, "How are relations connected to one another and how might those connections affect actions?"

Galileo and Newton had this advantage over Weber. They had a geometry to picture their phenomena. That geometry came down to them from Euclid. They used it and built on it. At Weber's time there was no geometry that could be used to picture social relations. Theorists now are more fortunate. Unlike Weber or Marx, there is network geometry. Today, social networks is a well-developed area of study (Freeman, 2004). Up till now, however, networks have been more often used to picture observed paths of activity, but less often used to represent theoretical constructs. Given the many successful uses of geometry for representing phenomena in physics, chemistry and, more recently in biology (Watson, 1969), the authors suggest that any social theory that does not have a mode for representing social relations and social structures is incomplete. (See Chap. 7 of this book.)

A sense of how Weber developed the array of formulations used in his historical investigations is given by his definition of ideal type.

> An ideal type is formed by the one-sided *accentuation* of one or more points of view and by the synthesis of a great many diffuse, discrete, more or less present and occasionally absent *concrete individual* phenomena, which are arranged according to those one-sidedly emphasized viewpoints into a unified *analytical* construct. (1904/1949, p. 90 [italics original])

The strength of this definition lies in its emphasis on one-sidedness. For example, looking back at the paragraph from Marx on consciousness and historical change, there is a series of embryonic ideal types. Are ideal types abstract concepts? Or are they a kind of half-way house, not fully abstract and not entirely concrete? The authors think the latter.

Could a worldview as pluralistic as Weber's be fashioned into theory as determinist as Marx's? No doubt that it could, but only if its concept formation were more systematic. There are ideas in Weber worthy of incorporation into more formal theory. For example, borrowing from Marx and generalizing, Weber coined the separation of the warrior from the means of violence, the separation of the administrator from the means of administration and even the separation of the laity from the means of salvation. Furthermore, Weber saw mobility in bureaucratic hierarchy as importantly related to power centralization. Insofar as a more formal theory of history is concerned, the problem is not whether idealism and materialism can be mixed systematically. The problem is whether such a broad-gauge theory can be more precisely formulated.

More Recent Worldviews

Where should scholars look to find worldviews more recent than those of Marx and Weber? Just after midcentury, Merton warned against the "balkanization of sociology" (1957, p. 51), a process that has increasingly divided the field into specialized areas. Currently the American Sociological Association has 53 Sections, each representing an area of specialization. Each Section organizes sessions in which papers reporting recent research are presented. Given this cacophony of voices, how can a selection of worldviews avoid being arbitrary?

Seeking a system for selection, the authors choose a two-step process. The first step was to select for the worldviews of sociologists who had been president of the American Sociological Association in the second half of the twentieth century. Within that grouping the selection was narrowed to those most interested in theory. The result was two notables, George Homans (1964) and James Coleman (1992). Limiting selection to past ASA presidents has the advantage that each published a 'presidential paper' in *The American Sociological Review* (Coleman, 1993; Homans, 1964) setting forth their views. Those papers became our starting point for uncovering the worldviews of each.

George Homans

There is a certain irony in the election of George Homans as president of the American Sociological Society because central to his worldview is the claim that there are no sociological propositions. As he asserts:

> it must be that the general explanatory principles even of sociology are not sociological, as the functionalists would have them be, but psychological propositions about the behavior of men, not about the behavior of societies. (Homans, 1964, p. 815)

Though Homans asserted that there are no sociological principles, he did not see that position as 'reductionist'.

> For reduction implies that there are general sociological propositions that can then be reduced to psychological ones. I now suspect that there are no general sociological propositions. (1964, p. 817)

According to Homans, sociologists should adopt the "'covering law' view of theory" (1986, p. xiii). "The only inescapable office of theory is to explain" (1964, p. 811). Theory consists of general propositions that enter into many deductive systems (1964, p. 812). In considering what a theory is, Homans based his description on a study "of some, not all, of the philosophers of science" (1986, p. xix). From that study he adopted the 'covering law' view of theory (1986, p. xix). Though Homans named no philosopher of science, from the term 'covering law' the reader will undoubtedly recognize Carl Hempel's deductive-nomological paradigm. Homans adds, "The explanation of a phenomenon is the theory of the phenomenon" (1964, p. 812).

The covering laws that Homans accepts are 'the laws of behavioral psychology' (1986, p. xiii). In Homans' "Fundamental Social Processes" a list of five propositions is given as 'basic' (1967, pp. 32–41). Here are the first two:

> "PROPOSITION I. *The more often a person's activity is rewarded, the more likely he is to perform that activity.* (1967, p. 33 [italics original])
>
> "PROPOSITION II. *If in the past the occurrence of a particular stimulus, or set of stimuli, has been the occasion in which a person's activity has been rewarded, then the more similar the present stimuli are to the past ones, the*

> *more likely the person is to perform the activity or some similar activity, now.* (1967, pp. 33–34 [italics original])

Of the five propositions, three employ the phrase "more likely" as a connective. Perhaps the five propositions were intended to serve as major premises in deductive explanations. If so, as statements of likelihoods, they cannot be so employed. Major premises in deductive syllogisms are absolute, not probable. They cannot have exceptions. That they cannot was clearly seen by Hempel (1965, p. 55ff).

It seems that there is a fundamental contradiction in Homans' theorizing. On the one hand, theory must be deductive. On the other hand, the only propositions Homans will accept, those of behavioral psychology, are probable, not absolute. Therefore, they cannot be major premises in deduction.

There is a second difficulty in Homans' sociology, the claim that there are no sociological propositions. His claim was universal, that there could not *ever* be sociological propositions. In the years since Homans, however, a number of exchange theories have been developed, all of which find the determination of exchange ratios in social structural conditions. For example, the authors of this book published a paper in which the predictions of fully ten exchange theories were compared (Willer & Emanuelson, 2008). All ten theories predicted the distributions of exchange ratios from structural conditions. An array of structures is studied. None of the structural conditions employed in predicting exchange ratios were reducible to psychological conditions.

Interestingly, all ten theories had some success, even Power-Dependence Theory. The success of power-dependence theory was surprising because that theory began as an exercise in reduction—with the aim of reducing structural to operant psychological conditions (Emerson, 1972a, 1972b). Having failed in that enterprise, the theory did an unacknowledged reversal of field to predict exchange ratios from structural conditions.

Tests of Network Exchange Theory employed by the authors of this book have been published in 18 papers including 4 successful international replications. In light of the record of successful research predicting human behavior from structural conditions, Homans' claim for the primacy of psychology is now fully falsified.

James Coleman

In the first sentence of James Coleman's 900+ page *Foundations of Social Theory* he states: "A central problem in social science is that of accounting for the functioning of some kind of social system" (1990, p. 1). Coleman's sociology is not at all like that of Homans'. The central problem of social science, as Coleman defines it, is located beyond the individual, a level not recognized by Homans. Coleman adds the following:

> Yet in most social research, observations are not made on the system as a whole, but on some part of it. In fact, a natural unit of observation is the individual person;...the dependence on individual-level data—most often interviews...has increased greatly...This has led to a widening gap between theory and research. (Coleman, 1990, p. 1)

Coleman's interest in theory is quite broad, ranging from the classical theory of Max Weber to contemporary rational choice. Insofar as research is concerned, however, his interests do not stray far from interview-based survey research.

Coleman recognized that contemporary social theory is rarely, if ever, concerned with the attitudes of individuals and that classical theory, whether it be that of Marx, Weber, Durkheim or Simmel, is not at all concerned with individual attitudes. He also undoubtedly knew that survey research could only collect information on the attitudes of and facts about individuals. He knew that because it has long been understood that survey research rips people out of the social structures in which they act. It follows that there is a gap between social theory and survey research, that may well be unbridgeable. In Colman's worldview, the gap not bridged came under the terms 'aggregation problem' and 'micro-macro problem' (1990, p. 769ff).

While Coleman is noted for his individualistic view of society, explaining events at the societal level is central to his theory. In his worldview, changes in society—in the macrostructure—are produced by the "somehow aggregated" actions of individual people—at the micro level. He agrees with Weber that "for sociological purposes there is no such thing as a collective personality which 'acts'" (1918/1968, p. 14). Therefore, explanation of structural change must dip down to the micro level and subsequently reconnect to the macro level.

For dynamics at the micro level, Coleman proposes to "use the conception of rationality employed in economics" (1990, p. 14). The rational actor is borrowed from economics as is the perfect social system which is the economist's perfect market (1990, p. 719). It is the task of the perfect social system to aggregate actions while moving from micro to macro levels. Then the macro behavior of the social system is explained by dynamics at the micro level—by the actions of its parts (1990, p. 2). He calls this approach "the internal analysis of system behavior" (1990, p. 2). This approach will provide "an understanding of the system behavior which a purely system-level explanation does not" (1990, p. 4).

To illustrate his approach, Coleman turns to a critical analysis of Weber's *Protestant Ethic and the Spirit of Capitalism* (1904/2002). For that analysis he introduces the diagram given in Fig. 5.4 that has come to be called the Coleman boat.

The vector across the top connecting Protestant religious doctrine and capitalism is not the explanation of the change from feudalism to capitalism, but the observed change at the societal level to be explained. The explanation is found by tracing Vectors 1, 2 and 3. Vector 1 traces from societal level doctrine to microlevel values inculcated by individual actors, who, tracing from left to right across Vector 2, are hard-working ascetic capitalists and workers. Vector 3 is intended to aggregate microlevel actions to produce capitalism. What is missing in Weber's historical explanation, according to Coleman, is how the actions of the ascetic rational actors came together, which is to say were aggregated, to form capitalist organizations in capitalist society (1990, p. 8). Weber failed to explain that aggregation—and so did Coleman (1990, p. 769ff).

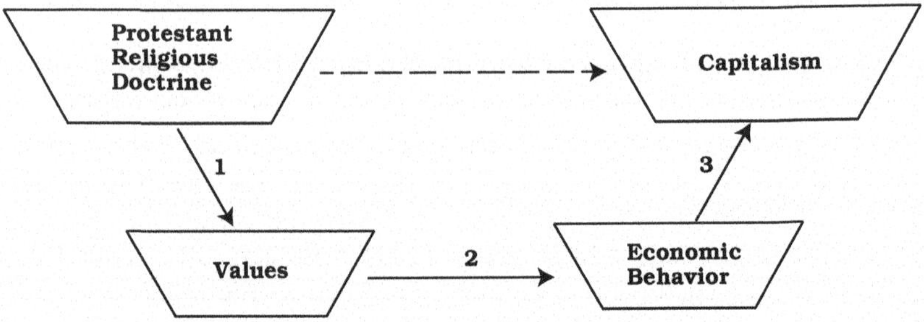

Fig. 5.4 Coleman's diagram for the effects of religious doctrine on economic organization

Coleman summarized his worldview as follows;

> What is necessary to account for the growth or occurrence of any social organization, whether capitalist economic organization or something else, is how the structure of positions constituting the organization comes into being, how persons who come to occupy each of the positions in the organization are motivated to do so, and how this interdependent system of incentives is sustainable. These are the central problems of the analysis of social organization. Marx's analysis of the emergence of capitalism from feudalism, polemically marred though it was, came closer to this than did Weber's analysis in *The Protestant Ethic*. (1990, p. 9)

There is one methodology in contemporary sociology where there is no unbridgeable gap between social theory and research and that is the theory-driven experiment. As already explained in Chap. 4, theory-driven experiments are laboratory investigations *designed* by theory to *test* theory. In Chap. 8 of this book, the authors explain how to use theory to build theory-driven experiments. Chapter 7 emphasizes that theory-driven experiments are not restricted to "small group studies". That is to say, the implications of sociological theory-driven experiments are not limited to small groups as Weber (1918/1968, p. 10) and many others have asserted. The size of social structure to which experimentally tested theory can be applied is wholly determined by the scope of the theory tested. As we will show in Chap. 8, for theory-driven experimentation there is no micro-macro problem. There is none because the analysis begins with actors in structures that are carried across Vectors 1, 2 and 3 to give explanation at the end of Vector 3.

We do not agree with Coleman that Marx's analysis of the development of capitalism is polemically marred. As discussed above, Marx's explanation is found toward the end of *Capital* volume one under the topic 'primitive accumulation'. By primitive accumulation Marx meant the original separation of people from their means of production. Once separated from the means they formed the propertyless proletariat, a class necessary for capitalist production. Our explanation begins with the expulsion of agricultural workers from the land.

Figure 5.5 is a model for the overly full labor market which, as shown in the boat of Fig. 5.6, is at the end of Vector 1 and beginning of Vector 2. Moving across Vector 2 is the competition of workers for positions that cheapens labor driving it to subsistence. Sliding up Vector 3 gives a fully developed propertyless working class to fit into capitalist organizations.

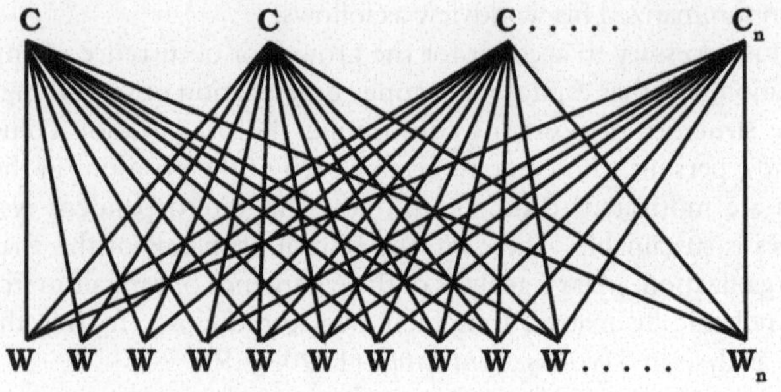

Fig. 5.5 An overfull labor market

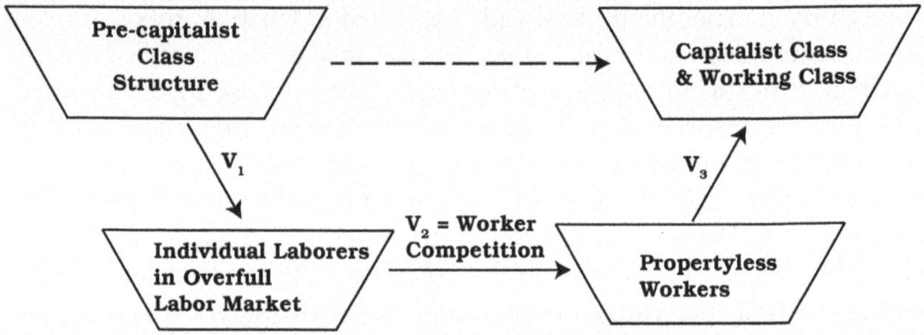

Fig. 5.6 Formation of capitalist classes in Coleman's Boat

Not included in this explanation sketch are the equations for the motives of the newly propertyless that produce the decline in wages. Those equations are introduced in Chap. 7 and experimental results supporting the dynamic suppression of wages are given in Chap. 8.

Also not included in this explanation-sketch are a number of further social structures and their structural dynamics involved in the rise of capitalism. Some of those structures include the newly emerging state, newly invented private property system and the structure of mobile capital. Does this complexity mean that explanation is impossible? To the contrary, Chap. 9 shows how to employ a theory toolbox to build an array of models that, once linked, provide a comprehensive modeling of this and other social changes. Still, modeling primitive accumulation is a good beginning point for an explanation of the emergence of capitalism.

Since there is a methodology that can close the gap between theory and research, why did Coleman not employ it? There is no doubt that

Coleman's scholarship is impressive. For all of his broad-based knowledge of the social sciences, however, judging from the indices of *Foundations of Social Theory* and "Social Theory, Social Research and a Theory of Action" (1986), he had limited exposure to experimental research in sociology. Perhaps it is in theory-driven experimentation, not in survey research, that Coleman would have found the solution to his micro-macro problem.

Conclusion: Worldviews

What has been learned from this survey of worldviews? First and foremost, it is evident that scientific theories that do not have a mode of representing their phenomenon are incomplete. The authors feel that that statement is quite general and by no means limited to social theory. In antiquity, not Aristotle, but Archimedes (1897) used geometric representations to solve empirical problems theoretically. When he was murdered by a Roman soldier, the science of antiquity died with him. Almost two millennia were to pass before Galileo revived the use of scientific theory.

Second and equally important, none of the worldviews of social theorists reviewed here employed a geometry to represent social phenomena. Of course, Coleman had his boat, but it represented the thinking of the theorist, not the phenomena under investigation. Since none represented phenomena, in theorizing, all were more akin to Aristotle than to Galileo or any more modern scientist. It would be unkind to suggest that there is a two-millennial lag in the development of social theory. But reading Homans is just a bit like reading Aristotle.

It has been suggested that developing social theory has proved particularly difficult because its subject matter is not about objects but about events—such as social action. Its subject is not solid like a rock, but ephemeral like a whisp of smoke. But this is wrong. Modern science began with Galileo's study, not of objects that moved, but of the events of motion itself. *Furthermore, when a geometric representation is drawn, events are represented in the model as objects.* That events are represented by geometric objects is as true for sociology as it is for any of the more successful sciences.

There is a stream of thought in sociology that sees economic theory as an ideal toward which sociological theory should strive. This stream of thought has diverse tributaries. For Talcott Parsons it meant that, just as economics was grounded on economic man, sociology should be grounded on normative man (Parson & Shils, 1951/2001). For Coleman it meant that sociology should borrow formulations such as economic man and the equilibrated market from economics. But normative man has no dynamic. It is like an insect frozen in amber. And social structures from chiefdoms to bureaucracies are not structured like markets and, as a consequence, each has its own dynamic. Economic theory is not the starting point for sociological theory.

If a mode of representing social phenomenon is to be the starting point, social theorists have little to learn from economic theory. The student of social phenomena knows full well that social structures take many forms, none of which is an equilibrated market. Nor is it given *a priori* that any social structure, other than a perfect market, equilibrates. To the contrary, it is the task of social theorizing to discover, which is to say invent, the dynamics of each social structure that can be drawn.

In concluding this chapter something more should be said about rationality. Two types of rationality have been identified, strategic and parametric. The two were differentiated by Von Neumann and Morgenstern just before the mid-twentieth century (1944, p. 11ff). Subsequently, Schelling (1970, p. 86) defined strategic rationality as occurring when "each player's best choice of action depends on the action he expects the other to take". By contrast, parametric rationality optimizes between given alternatives (Elster, 1986, p. 7). In fact, this differentiation was anticipated by Weber. "The economic activity of an individual is social only if it takes into account the behavior of someone else" (1918/1968, p. 22). Human activity in a perfect market is parametric. Social actions in social structures are strategic. It makes sense to turn from the economic man's parametric rationality to Game Theory's strategic rationality when formulating the social actor.

Today, the point of departure for social theory should be to develop modes of representation for social structures that can then be filled with strategically rational social actors from which the dynamics of those structures can be inferred. Developing modes of representation begins in Chap. 6.

References

Archimedes. (1897). *The works of Archimedes* (T. L. Heath, Ed.). Dover Publications. (Original work published ca. 232 B.C.E.).

Aristotle. (1961). *Physics* (R. Hope, Trans.). The University of Nebraska Press. (Original work published ca. 330 BCE).

Coleman, J. S. (1986). Social theory, social research, and a theory of action. *The American Journal of Sociology, 91*(6), 1309–1335.

Coleman, J. S. (1990). *Foundations of social theory.* The Belknap Press of Harvard University Press.

Coleman, J. S. (1992). The rational reconstruction of society. *Annual Sociological Review, 58*(1), 1–15.

Coleman, J. S. (1993). The rational reconstruction of society: 1992 Presidential Address. *American Sociological Review, 58*(1), 1–15.

Einstein, A. (1954). *Ideas and opinions.* Crown Publishers.

Elster, J. (1986). *Rational choice.* New York University Press.

Emerson, R. M. (1972a). Exchange theory part I: A psychological basis for social exchange. In J. Berger, M. Zelditch Jr., & B. Anderson (Eds.), *Sociological theories in progress* (Vol. 3). Houghton Mifflin.

Emerson, R. M. (1972b). Exchange theory part II: A psychological basis for social exchange. In J. Berger, M. Zelditch Jr., & B. Anderson (Eds.), *Sociological theories in progress* (Vol. 3). Houghton Mifflin.

Freeman, L. C. (2004). *The development of social network analysis: A study in the sociology of science.* Empirical Press.

Galileo, G. (1954). *Dialogues concerning two new sciences* (H. Crew & A. Desalvio, Trans.). Dover Press. (Original work published 1638).

Hempel, C. G. (1965). *Aspects of scientific explanation and other essays in the philosophy of science.* The Free Press.

Homans, G. C. (1964). Bringing men back in. *American Sociological Review, 29*(5), 809–818.

Homans, G. C. (1967). Fundamental social processes. In N. J. Smelser (Ed.), *Sociology: An introduction.* John Wiley & Sons, Inc.

Homans, G. C. (1986). Fifty years of sociology. *Annual Review of Sociology, 12*, xii–xxx.

Marx, K. (1967). *Capital.* International Publishers. (Original work published 1867).

Marx, K. (2009). *A contribution to the critique of political economy* (N. I. Stone, Trans.). Lector House. (Original work published 1859).

Merton, R. K. (1957). *Social theory and social structures.* Free Press. (Original work published 1949).

Newton, I. (1966a). *Mathematical principles of natural philosophy and his system of the world: The motion of bodies* (A. Motte, Trans. 1729, Vol. 1). University of California Press. (Original work published 1686).

Newton, I. (1966b). *Mathematical principles of natural philosophy and his system of the world: The system of the world* (A. Motte, Trans., 1729, Vol. 2). University of California Press. (Original work published 1687).

Parsons, T., & Shils, T. (2001). *Toward a general theory of action.* Transaction Publishers. (Original work published 1951).

Schelling, T. (1970). *The strategy of conflict*. Harvard University Press.
Von Neumann, J., & Morgenstern, O. (1944). *Theory of games and economic behavior*. Princeton University Press.
Watson, J. D. (1969). *The double helix: A personal account of the discovery of the structure of DNA*. Signet.
Weber, M. (1946). The social psychology of world religions. In H. Gerth & C. W. Mills (Trans. & Eds.), *From Max Weber: Essays in sociology*. Oxford University Press. (Original work published 1922).
Weber, M. (1949). *The methodology of the social sciences* (E. A. Shils & H. A. Finch, Trans.). The Free Press. (Original work published 1904).
Weber, M. (1968). *Economy and society* (Vol. 1). University of California Press. (Original work published 1918).
Weber, M. (2002). *The Protestant ethic and the spirit of capitalism* (S. Kalberg, Trans.). Roxbury Publishing Company. (Original work published 1904).
Willer, D., & Emanuelson, P. (2008). Testing ten theories. *Journal of Mathematical Sociology, 32*, 165–203.

CHAPTER 6

Contemporary Theories and Schools of Thought

A science's success can be understood as a function of the breadth and precision of its explanations. Since there is no inductive methodology, seeking general knowledge, science aims to build theory. Ideally, theory takes a form that can design experiments because, as discussed in Chap. 4, the interplay of experiment and theory is at the heart of the falsification cycle. That cycle is a process whereby theory is tested, and results guide modification to the theory, which is then tested in an ongoing cycle until consistent support of tests occurs.

A fully formed theory should include a worldview, a modeling procedure and mechanisms/laws. As will be seen in this chapter, however, very few social theories have a modeling procedure. Nevertheless, many have mechanisms or laws or both. But building a fully developed theory is difficult. Proto-theories are incomplete theories that are developed only in part. As seen here, they can take many forms. Some have a diagramming procedure but use an imprecise mechanism. Others provide a coherent worldview but lack formal models. Several include robust mechanisms but no diagramming procedure.

Proto-theories significantly contribute to scientific knowledge as stepping stones toward fully developed theory. However, because they are not complete, their contribution is limited. What follows is a discussion of four contemporary proto-theories. Most include one or more components of the form proposed here. The intention is not to review any of them in

D. Willer, P. Emanuelson, *The Logic of Theoretical Sociology*,
https://doi.org/10.1007/978-3-031-91709-7_6

their entirety, but to provide enough detail to highlight each theory's form and discuss its implications. In addition, the sequence of changes in Power-Dependence Theory, changes that are not a falsification cycle are reviewed. That review uncovers a number of weaknesses of sociology as a science.

Balance Theory: Modeling Sentiments

Balance Theory (1946, 1958/2015) is worthy of study because it is nearly unique among social theories as having both a modeling procedure and a mechanism. Heider introduces his Balance Theory by defining sentiments as positive or negative valuations of persons or objects (1958, p. 77). Balance theory models are diagrams like those shown in Fig. 6.1. The purpose of the diagrams is to trace the implications of positive and negative valuations and to determine when the valuations are in balance or harmony and when they are not. If not, what are the consequences of that imbalance? Following Heider's initial 1946 statement of the theory, an important paper by Cartwright and Harary (1956) offered a generalization. Both will be reviewed here.

Let us begin by describing the six triangle diagrams of Fig. 6.1 as introduced by Heider. At the base of each triangle are the nodes 'p' and 'o', person and other. The valuations are p's valuation of o, p's valuation of 'x' an attitude object, and p's inference concerning o's valuation of x. The x

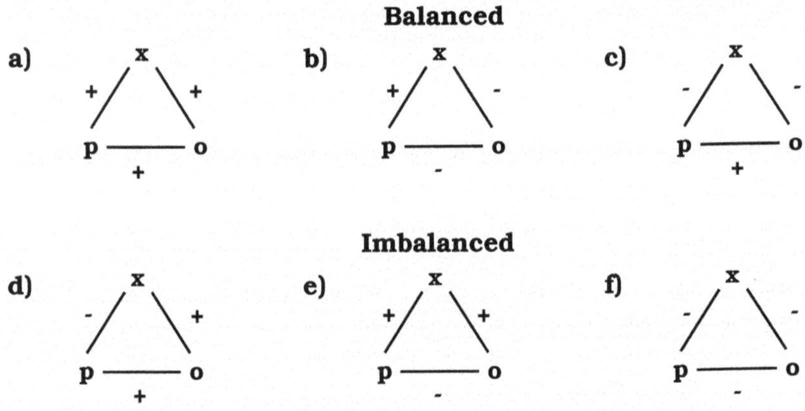

Fig. 6.1 Heider's six diagrams of cognitive balance and imbalance

node can be valued positively or negatively by p and o but does not, itself, value any other node.

It must be emphasized that the diagrams are drawn from the point of view of p. Thus, in the diagrams of Fig. 6.1, the signs of the p – x and o – x relations are evaluations by p of x and p's inference concerning o's valuation of x. The signed line between p and o is p's evaluation of o. In this, Heider's original formulation, all evaluations are from the point of view of p. Later, in their generalization, Cartwright and Harary draw directed lines, one connecting p to o giving p's valuation of o and one from o to p giving o's evaluation of p. Now consider the 'balance' of the figures of Fig. 6.1. Assume that a positive sign is + 1 and a negative sign is −1. Now multiply the signs of the top three networks. The result is +1 because (+ 1) (+ 1) (+ 1) = +1, (+ 1) (−1) (−1) = +1 and (−1) (+1) (−1) = + 1. For all three, +1 indicates that the network is balanced. 'Balanced' means that the parts fit together such that there is no pressure to change. For example, when all signs are positive, p and o like each other, and both like x: there is a sense an internal consistency. When p and o dislike each other and one of the two dislikes x, again, a consistency is sensed. For all three, balance implies that there is no pressure to change. Importantly, there is a good fit between the sign multiplication mechanism and our intuitions.

Conversely, multiplying the signs of each of the bottom three gives −1, indicating that they are imbalanced. In 'd', for example, p and o like each other, but p dislikes the x that is liked by o. For Heider, when a diagram is imbalanced, there is a pressure toward balance. In 'd', there are three ways that it can come to be balanced: (1) p can come to like x, (2) o can come to dislike x or (3) p and o can come to dislike each other. Similarly, networks 'e' and 'f' can become balanced in any of three ways. Significantly, Balance Theory does not predict which of the three changes will happen. It does, however, predict that, having come to balance, there is no further pressure to change.

An important generalization of Heider's theory was offered by Cartwright and Harary (1956). Following a brief review of Heider and parallel work by Newcomb (1953), they conclude that the tendency toward balance "is a significant determinant of cognitive organization" (p. 280). They transfer the "point of view" of the diagrams from p to outside the structure by introducing the idea of a digraph. In a digraph, points are connected by directed lines between nodes. This transfer is seen by comparing Heider's 'a' structure to Cartwright and Harary's balanced structure given in Fig. 6.2. Now, p and o are both evaluating nodes.

Fig. 6.2 Cartwright & Harary's generalization of Heider's graph

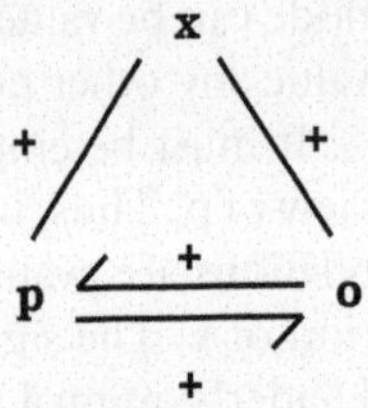

Therefore, there are two directed lines between them, one representing p's evaluation of o and the other o's evaluation of p. There are directed lines connecting p to x and o to x, but no corresponding x to p or x to o directed lines. This is because x is an attitude object, not an evaluator. Then Cartwright and Harary introduce the multiplying sign rule already given above and proceed to demonstrate that their formulations are "a rigorous generalization of Heider's concept of balance" (p. 283). They demonstrate that this elegantly simple idea of balance can be extended to any number and size of signed graphs.

Because there are too few examples of cumulative development of theory, what Cartwright, Harary and Heider, before them, have done is highly commendable. This kind of work marks off the building of scientific theory from lesser work. Nevertheless, Balance Theory still needs to be completed because it is indeterminant insofar as the transition from imbalance to balance is concerned. When a network is not in balance, one of three transitions will produce balance. Heider gives no guidance, however, as to which of the three will occur. Nor do Cartwright and Harary. The theory is incomplete, an incompleteness that sharply limits its value as a scientific theory.

Game Theory

Unlike Balance Theory, Game Theory is not a single theory, but what could be called a family of theories. The family of theories that constitute Game Theory is distinguished by the sophistication of its mathematical and logical formulations and by its focus on issues of cooperation and conflict. Its point of departure can be traced to von Neumann and Morgenstern's *Theory of Games and Economic Behavior* (1944). The economic behavior at the center of their interest involved not the parametric decision-making where economics had previously focused, but on strategic decision-making. According to Elster, "In a parametric decision the agent faces external constraints that are in some sense given or parametric.... A strategic situation is characterized by interdependence of decisions" (1986, p. 7). The parametric–strategic distinction was anticipated by Weber. For Weber, economic action is parametric because the market fixes prices, but social action involves interdependent actions (1918/1968, p. 4ff). Given this common point of departure, one might expect to see Game Theory at the core of sociological theory, but it is not. Why it is not is an issue which will be addressed.

Reviewing a few games and their solutions will give the reader a feeling of how Game Theory approaches cooperation problems. Though its payoff matrix was first published by Flood (1952), Albert Tucker popularized the Prisoner's Dilemma (hereafter P/D) narrative associated with that matrix (Poundstone, 1993). Here is the narrative. Two people charged with a crime are held separately by the police. Each is told that the police already have sufficient evidence to send them both to prison for one year. If they both confess, however, they will be sentenced to two years. But if only one confesses, that one will go free (the zero payoff) and the other will be sentenced to three years.

The game associated with this narrative is given in Fig. 6.3. The payoffs to Player 1 are to the left of each box, while the payoffs to Player 2 are to the right. Player 1 plays row while player 2 plays column. The "C" play is to cooperate, and the "D" play is to defect. The two players play without knowledge of the other's play. As can be seen, when both cooperate, the '−1' is one year in prison whereas when both defect −2 is two years in prison, and similarly for the off-diagonal payoffs. The ordered payoffs can

Fig. 6.3 One State of the Prisoner's Dilemma

		Player 2	
		C	**D**
Player 1	**C**	**-1, -1**	**-3, 0**
	D	**0, -3**	**-2, -2**

be abbreviated as DC > CC > DD > CD. Furthermore, each payoff has a name: DC is temptation (T) > CC is reward (R) > DD is penalty (P) > CD is sucker (S). Then the payoffs can be further abbreviated as T > R >P >S. The game is a dilemma because its dominant strategy equilibrium—the Nash solution—is for both players to defect, to play D. Yet playing C, joint cooperation gives the better payoff for both.

Figure 6.4 shows three notable games: P/D, Chicken and Stag Hunt. In comparing P/D in Figs. 6.3 and 6.4, the first is converted into the second by adding 3 to each payoff; the game is unaffected by adding a constant. The conversion facilitates the comparison of P/D with the other two games. The narrative for the chicken game is not about the bird, but, instead, pictures two juveniles driving toward each other at high speed. Cooperation is to swerve just prior to the crash, and defection is to continue straight. If both "chicken out", they live for a payoff of 2 each. If one chickens out, the other wins. If neither "chickens out", both defect and die for a zero payoff.

The third game is the stag hunt. This is the narrative for the game. If both players cooperate, they get a stag for a payoff of 3 each. If either player is distracted by a passing rabbit, that player gets 2, and the other gets zero. If both are distracted by the rabbit, they split it for a payoff of 1.

The application of these three games, and others like them, is not restricted to the narrative with which they are typically identified. For example, in the next chapter, the application of strategic analysis uncovers the P/D dilemma in power structures where it is otherwise hidden.

Nevertheless, applications of Game Theory are notable for its mixture of mathematical rigor and empirical content. For example, applying Game Theory to international conflict mixes empirical assessment of military strength with abstract game theoretic calculations.

At various times over the last half-century, Game Theory has been touted as *the* new social theory. Beyond its impressive mathematical core, however, no theory of social action, social relations, or social structure has been built on Game Theory. Until then, Game Theory will not be the basis for social theory. On the other hand, Game Theory has much to say about conflict. That it does is both a virtue and a curse. It is virtue in that its applications to conflict remain the most impressive of its derivations. It is a curse because, through long-term close association with the military, Game Theory has forfeited any claim it might have had to the value neutrality that is essential to being a social theory.

a) PD Game

Player 1 \ Player 2	C	D
C	2, 2	0, 3
D	3, 0	1, 1

b) Chicken Game

Player 1 \ Player 2	C	D
C	2, 2	1, 3
D	3, 1	0, 0

c) Stage Hunt

Player 1 \ Player 2	C	D
C	3, 3	0, 2
D	2, 0	1, 1

Fig. 6.4 Three notable games

Rational Choice Theory

While Game Theory is not a unitary theory but a family of theories, rational choice theory is not a family of theories but an orienting perspective. It is a viewpoint toward what social theory should be. Here is Coleman and Fararo's statement of the goal of rational choice:

1. The set of phenomena to be explained by the theory is the behavior of social systems (large or small) and not the behavior of individuals.
2. Explanation of the behavior of social systems requires explanation in terms of the behaviors of actors in the system (1992, p. ix).

Rational Choice Theory combines holism at the structural level and individualism at the actor level into a new orientation Udehn calls 'structural individualism' (2001, p. 318). Since its ultimate goal is explanation at the system or structural level, the label 'rational choice' may be misleading. Or perhaps not, because much of what counts as theorizing will necessarily occur at the actor level.

Central to rational choice's orientation is the idea that the actor-level and the system-level are somehow separated. If a rational choice theory is to be successful, it must deal with that separation. More than the behavior of actors needs to be explained. Formulations are required to cover the "transitions between the level of social system behavior and the level of behavior of individual actors" (Coleman & Fararo, 1992, p. ix). How the two levels are thought to be separated is displayed in Fig. 6.5, which has come to be called the Coleman boat.

As explained in Chap. 5, Coleman's boat is not a diagram of a social structure. To the contrary, it is intended to illustrate how the rational choice theorist must think to solve problems in theory. The directed line across the top is the macro social change to be explained. Directed line number one leads from the macro to the micro level, directed line two is

Fig. 6.5 Coleman's Boat

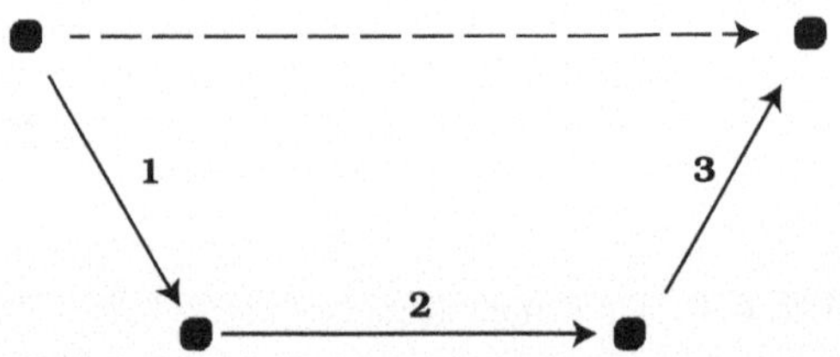

the micro-dynamics and directed line three leads back from the micro to the macro level. Directed lines one and three stand for the formulations needed to cover the transitions from macro to micro and back again. Therefore, a complete rational choice theory contains three theories (macro → micro, micro-dynamics, micro → macro), all articulated with each other. Building theory is a challenging task. Building three articulated theories may well be an impossible task.

It is the conviction of the authors of this book that building three articulated theories is not now needed to explain the dynamics of large structures, nor will it ever be required. The evidence supporting that conviction consists of these points. First, the techniques of scientific theory building have already been employed in building Elementary Theory. That theory and its methods will be introduced and explained in Chap. 7. Second, how experiments are constructed to test theory is shown in Chap. 8. Examples employed in that chapter include experimental designs used to test Elementary Theory, some cross-nationally. Third, Chap. 9 gives three example applications of Elementary Theory to macrostructures. At no point in these expositions is anything like the theorizing demanded of rational choice or the Coleman boat to be found.

Furthermore, there are many examples of theory-driven work in the rational choice tradition in which nothing like a Coleman boat is to be found. An example is Mancur Olson's influential *The Logic of Collective Action*. Olson begins by pointing out that diverse schools of thought, as distinct as Marxism on the one hand and "group theory" on the other, take it as given that groups will act collectively to serve the interests of their members (Olson, 1965, p. 1). Olson asserts to the contrary that "*rational self-interested individuals will not act to achieve their common or group interests*" (1965, p. 2 [italics original]).

The kinds of organizations on which Olson focuses are those that are "expected to further the interests of their members" (1965, p. 6). Even when they do further members' interests, only a tiny proportion of those who could be members actually join. For example, only a small percent of the working class in the U.S. join labor unions, even though the ones who do join are generally better paid than those who do not. In Olson's terms, instead of joining, the majority "free ride" (1965, p. 76).

Do free riding in a collectivity and defection in a Prisoner's Dilemma appear to be the same thing? They should because they are. In this case, appearances are *not* deceiving. As shown in the later section on strategic analysis, defection in the P/D game and free riding in collective action are logically identical. Through that identity, micro and macro are connected without recourse to the three articulated theories of Coleman's boat.

Social Network Analysis

Social Network Analysis, like Game Theory, is distinguished by the sophistication of its mathematical and logical formulations. The good news is that its sophisticated formulations improve our understanding of social structures. The founders of Social Network Analysis knew full well that the social survey acts like a "meatgrinder, tearing the individual from his social context" (Freeman, 2004, p. 1). Social Network Analysis stood in the greatest possible contrast to the social survey in its focus, not on individuals, but on the connectivity of actors and the nature of the networks in which they are connected. Notably, many of those founders were mathematicians—thus the rigor of its foundations.

As late as the 1960s, Harary, Norman and Cartwright could justifiably assert that "The technical terminology employed in describing structures is meager; few concepts are defined rigorously...and detailed studies of structure, as such, are rather rare" (Harary et al., 1965). Their goal was to "present some mathematical theory about the abstract notion structure" (Harary et al. 1965, p. 2). Their work and the equally meritorious later work of Wasserman and Faust (1994) offer us a rich array of concepts connecting actors' actions to each other and to the structures in which action occurs. The bad news is that, despite bettering our understanding of actors in structures, Social Network Analysis has remained precisely that: a method of data analysis, not a theory of social structure. We will consider why it did not evolve into a theory of social structure upon concluding this section.

Nevertheless, the clarity of the presentation of technical terms in Social Network Analysis is refreshing. For example, a geodesic is the shortest path between two nodes (Wasserman & Faust, 1994, p. 110), and one meaning of the 'centrality of an actor' is 'degree', which is to say a simple count of the number of ties of that actor (p. 173). Looking at Fig. 6.6, by that definition, the node in the center of the star network is more central than the other nodes of that network. 'Ego density' is the ratio of the actor's degree to the maximum degree possible. For n_1 at the star's center, density is 1, while density is smaller than 1 for all other nodes. Centrality in the sense of 'closeness' is the total distance from *i* to all other nodes. It is the sum of all geodesists that include *i*. Finally, in the sense of 'betweenness', an actor is central if it is on the geodesic of many other actors. The

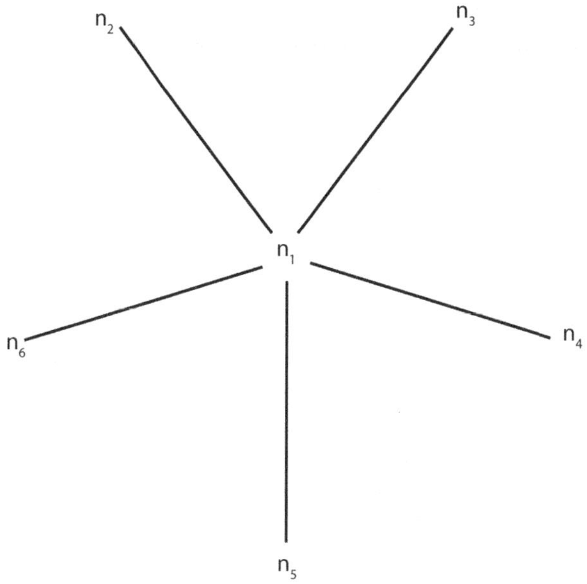

Fig. 6.6 A 6-node Star Network

clear and clean introduction of the concept of centrality is typical of the wider field of social networks.

Interestingly, the earliest form of network investigation was the study of kinship. In the nineteenth century, Lewis Henry Morgan (1871/1997) offered a scheme for multiple kinship structures. Notable work continued through the early to mid-twentieth century, notably by Radcliff-Brown (1922/2018), Malinowski (1927/2014, 1922/1984) and John Barnes (1971/2004). Nor have these been studies of static, lifeless structures. On the contrary, placement in the kinship structure of a traditional society is intimately related to social status and, thus to social influence. The study of kinship contains a critical dynamic. That dynamic is seen in Chap. 9 where the organizational structure of early chiefdoms is inferred from its lineage structure using structural formulations from Hage and Harary (1996) and linkages from status and number to influence through theory offered by Melamed and Savage (2013a, 2013b, 2016).

There is no doubt that Social Network Analysis has greatly benefited from its ties to an array of fields ranging from mathematics to electrical engineering. It is no longer true that the technical vocabulary of social structures is meager, nor is it ill-defined. Nor is it the case that there are few studies of social structures. Studies of social structures, from experimental to historical, are found throughout this book. Social Network Analysis benefits from being founded and carried forward by first-rate minds with a not insignificant injection of mathematicians. Why has Social Network Analysis *not produced* the social structural theory that many of us have sought?

The answer is two-fold. First, Social Network Analysis is a method of analyzing social structure. It is not a method for building theory about social structure. Crucially, that distinction means that social network analyses are structured by relations found in the data. Social Network Analysis does not impose order on its data. Therefore, it is not theory. Second and perhaps less important is the absence of a model for the actor. In Social Network Analysis, the actor is a node in the network. Nothing more. In Udehn's terms, Social Network Analysis is not structural individualism. It is structuralism.

A personal comment might clarify that last point. Social network scholars hold annual meetings. The first author of this book has enjoyed attending many. When chatting with Lin Freeman at one meeting, the author asked him why network analysts had yet to develop an actor model. He responded, "Who knows why people do what they do".

Power-Dependence Theory

For a scientific field to be progressive, its theoretic model must be falsifiable. For falsifiability, more is needed than for a model to be sufficiently precise that some of its predictions are found to be wrong. That kind of falsifiability only avoids regression. To be progressive, there must be a competing model that covers, without error, the predictions of the one being falsified, and offers new predictions over a broader scope. Finding a theory to be wrong is to be 'falsified'. Offering a new more accurate and broader scope theory is to be 'progressively falsified'. Both kinds of falsifiability are discussed in Popper (1934/1959) and Lakatos (1970). Importantly, progressive falsification is not possible if other criterion, such as the status of a theory's authors, guides its acceptance and rejection.

Exchange theories share the goal of relating social structure and exchange ratios. Power-Dependence Theory is unique among exchange theories in that "*social structure* is taken as the dependent variable"

(Emerson, 1972, p. 58 [italics original]). This form of Power-Dependence Theory applied operant psychology. Structure was the dependent variable because it is inferred from differential satiation effects on operant actors. If a centrally placed actor experiences favorable payoffs and actors connected to it experience unfavorable payoffs, that result is produced by the more frequent exchanging and, thus, more rapid satiation of the central actor. Rates of satiation are the independent variables, and structure, inferred from them, is the dependent variable.

However, Power-Dependence Theory only briefly maintained social structure as the dependent variable. Six years later, the reverse was proposed. "(T)herefore, we take power use as the main dependent variable" (Cook & Emerson, 1978, p. 724). This reformulation of the theory uses a rational actor—the operant actor is abandoned. Now, exchange ratios are determined by 'equal dependence' in the following way. Assume that actor A may exchange with B but has an alternative where A gains 12. By contrast, B has no alternative but to exchange with A. A's payoff when exchanging with B is RAB and AAB is A's best alternative—and similarly for B, the equation for equal dependence is as follows:

$$\text{RAB} - \text{AAB} = \text{RBA} - \text{ABA}$$
$$\text{Given RAB} - 12 = \text{RBA} - 0$$

The size of the pool to be divided is 24. Therefore, RAB + RBA = 24. Substituting and solving RAB = 18 and RBA = 6. This formulation for equal dependence predicts unilateral monopolies well.

However, it is unclear how the equal dependence equation is to be applied to the network in Fig. 6.7. In that network, the solid lines are relations in which 24 points are divided while, in the single dashed line, only 8 points are split. Since only the D position has two alternative high-profit relations, it would seem that it would be high power. However, the E

Fig. 6.7 Cook et al. (1983) Figure 1c, A 5-Person Network with Three Positions

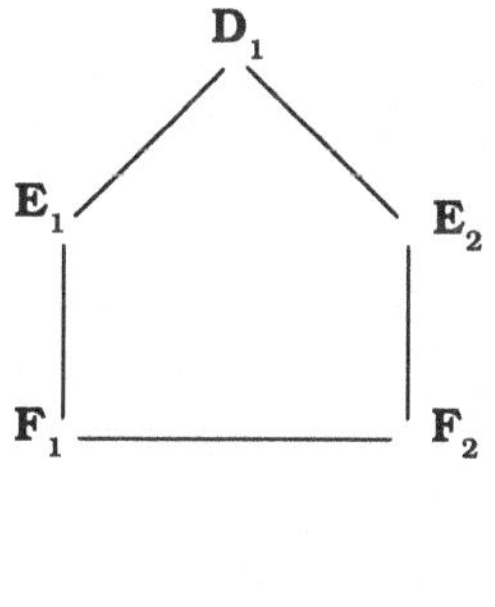

positions in the network of Fig. 6.7 gain the best payoffs and thus are higher in power than D or the Fs. It is unclear how that result would be calculated from the equal dependence equation above.

To differentiate high and low power positions, 'vulnerability', an entirely new formulation, was added to Power Dependence Theory. Vulnerability "locates the points of minimum dependence, equivalent to maximum network-wide power" (Cook et al., 1983, p. 300). Vulnerability is determined by removing nodes and their relations one at a time and evaluating the reduction of resource flow produced by that removal (Cook et al., 1983, p. 299). For example, removing D or either F does not reduce maximum flow. But removing either E does reduce maximum flow. Therefore, the Es are predicted to be high power. Whatever the value of Vulnerability is as a locator of power positions, it is difficult to understand what "network-wide power" is. In the resource pool networks that Cook and Emerson use, only adjacent positions divide pools with each other—nothing flows through the network. Therefore, all power is local. Power occurs only between adjacent positions. There is no network-wide power exercise.

Three years after Vulnerability was proposed, the first author of this book published a comment on Vulnerability. The network of Fig. 6.8 is the simplest network for which Vulnerability gives false predictions. Two exchanges are possible for the network as a whole.

Removing A1, B1 or B2 reduces the maximum to one exchange. Therefore, all three are identified by Vulnerability as high in power. However, power is an ordering relation. If A has power over B, then B is low in power and cannot exercise power over A. Just as two positions cannot exercise power over each other, it is logically and theoretically impossible for three positions such as A1, B1 and B2 to exercise power over each other. In their response, Cook et al. disagree and assert that that impossibility is "simply wrong from a power-dependence perspective" (Cook et al., 1986, p. 447). What is correct, however, they do not say.

Fig. 6.8 A Structure to Falsify Vulnerability (Willer, 1986, Figure 2)

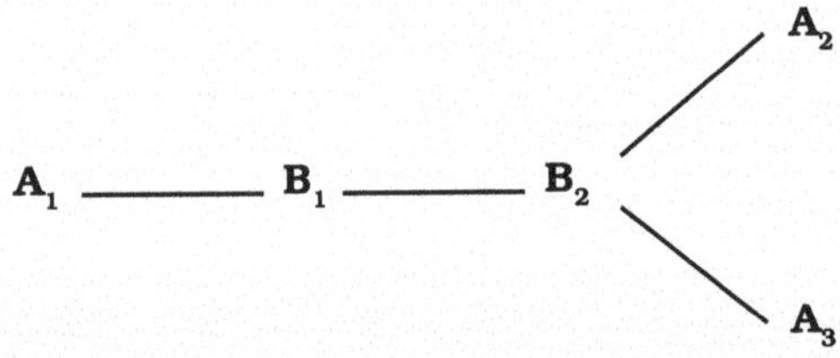

Cook et al. (1986) quite rightly point out that Willer did not offer an alternative to Vulnerability (1986, p. 445). That alternative had been submitted to the *American Journal of Sociology*—and subsequently rejected. That rejection is an unfortunate editorial choice, for it is regressive, not progressive science. Fortunately, the *American Sociological Review* published the paper authored by Markovsky et al. (1988) that offered the graph-theoretic power index (GPI) as an alternative to Vulnerability. In fact, GPI is a progressive falsification of Vulnerability.

Here is an example of the application of GPI to the network given in Fig. 6.8. GPI counts and sums nonintersecting paths through the network with odd-numbered paths being positive because they support power exercise and even-numbered paths being negative because they do not. Counting from A1, there is a single 1-path to B1, a single 2-path to B2, and a single nonintersecting 3-path to either A: GPI = 1 − 1 + 1 = 1. For A2, there are two 1-paths and a single 2-path: 2 − 1 = 1.

B2 has three nonintersecting 1-paths and a single 2-path: 3 − 1 = 2. A2 (or A3) has a single 1-path and one nonintersecting 2-path and one nonintersecting 3-path: 1 − 1 + 1 = 1. GPI = 1 for all positions but B2 for which GPI = 2. Only B2 is high power. Another part of the theoretic divides the network between the Bs where it predicts that no exchanges will occur. Experimental studies have supported these and other predictions of GPI.

The networks providing tests for GPI and PDT are not exchange networks for no resources move between positions. They are, instead, resource pool networks in which a pool of points, placed between each pair of nodes, is divided by joint agreement. Resource pool networks offer an advantage in that they can take any shape drawn by the theorist. Resource pool networks offer the disadvantage in that many of these shapes do not—perhaps could not—exist outside the laboratory. Figures 6.8 and 6.9 displays three weak power networks easily constructed in the laboratory for which no known example outside the laboratory exists.

For all of their limited—perhaps negligible—scope outside the laboratory, predicting exchange ratios in weak power networks posed an intellectual problem that attracted some of the best minds in sociology including Bonacich and Bienenstock (1995), Friedkin (1992, 1995), Yamaguchi and Cook (1990), Skvoretz and Willer (1993), Willer (1986) and Markovsky et al. (1993). The authors of this book tested ten theories (2008) for weak power resource pool networks. With one notable exception, all these theories are too complex for paper and pencil application: they must be applied through a computer program. Oddly, whereas

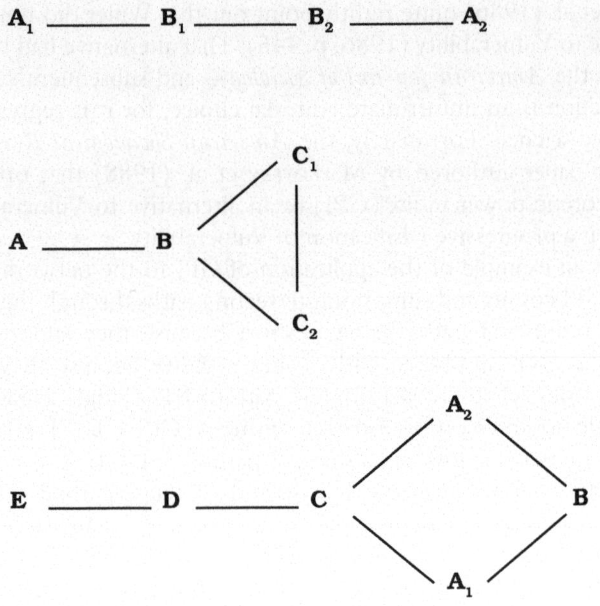

Fig. 6.9 Three weak power networks

resource pool networks are simple, theories testing the distribution of power in them are highly complex. Experimental tests found GPI to be the most precise while PDT scored ninth, second from the bottom of the ten theories tested.

This test of ten theories implies that sociology could be a progressive science. At the same time, the history of Power-Dependence Theory is disquieting. In every form Power-Dependence Theory has taken, it has been falsified. Yet, many still see it as a viable theory. That they do is no fault of the theory itself. The fault lies in the field of sociology. Its norms are not those of science. Until they are, sociology will not be progressive.

References

Barnes, J. A. (2004). *Three styles in the study of kinship*. Routledge. (Original work published 1971).

Bonacich, P., & Bienenstock, E. (1995). When rationality fails. *Rationality and Society, 7*, 293–320.

Cartwright, D., & Harary, F. (1956). Structural balance: A generalization of Heider's theory. *Psychological Review, 63*, 277–293.
Coleman, S. C., & Fararo, J. F. (1992). *Rational choice theory: Advocacy and critique*. Sage Publications.
Cook, K. S., & Emerson, R. M. (1978). Power, equity and commitment in exchange networks. *American Sociological Review, 43*, 721–739.
Cook, K. S., Emerson, R. M., Gilmore, M. R., & Yamagishi, T. (1983). The distribution of power in exchange networks: Theory and experimental results. *American Journal of Sociology, 89*, 275–305.
Cook, K. S., Gillmore, M. R., & Yamagishi, T. (1986). Point and line vulnerability as bases for predicting the distribution of power in exchange networks: Reply to Willer. *American Journal of Sociology, 92*, 445–448.
Elster, J. (1986). *Rational choice*. New York University Press.
Emerson, R. M. (1972). Exchange theory: A psychological basis for social exchange. In J. Berger, M. Zelditch Jr., & B. Anderson (Eds.), *Sociological theories in progress* (Vol. 3). Houghton Mifflin.
Flood, M. M. (1952, June). *Some experimental games* (Research Memorandum RM-789). RAND Corporation, June.
Freeman, L. C. (2004). *The development of social network analysis A study in the sociology of science*. Empirical Press.
Friedkin, N. (1992). An expected value model of social power: Predictions for selected exchange networks. *Social Networks, 14*, 213–230.
Friedkin, N. (1995). The incidence of exchange networks. *Social Psychology Quarterly, 58*, 213–222.
Hage, P., & Harary, F. (1996). *Island networks*. Cambridge University Press.
Harary, F., Norman, R. Z., & Cartwright, D. (1965). *Structure models: An introduction to the theory of directed graphs*. Wiley.
Heider, F. (1946). Attitudes and cognitive organization. *The Journal of Psychology: Interdisciplinary and Applied, 21*, 107–112. https://doi.org/10.1080/00223980.1946.9917275
Heider, F. (1958). *The psychology of interpersonal relations*. John Wiley & Sons.
Heider, F. (2015). *The psychology of interpersonal relations*. Martino Fine Books. (Original work published in 1958).
Lakatos, I. (1970). Falsification and the methodology of scientific research programmes. In I. Lakatos & A. Musgrave (Eds.), *Criticism and the growth of knowledge*. Cambridge University Press.
Malinowski, B. (1984). *Argonauts of the Western Pacific*. Waveland Inc. (Original work published 1922).
Malinowski, B. (2014). *The father in primitive psychology*. Morison Press. (Original work published 1927).

Markovsky, B., Skvoretz, J., Willer, D., & Erger, J. (1993). The seeds of weak power: An extension of network exchange theory. *American Sociological Review, 58*, 197–209.
Markovsky, B., Willer, D., & Patton, T. (1988). Power relations in exchange networks. *American Sociological Review, 53*, 220–236.
Melamed, D., & Savage, S. (2013a). Status, numbers and influence. *Social Forces, 91*, 1085–1104.
Melamed, D., & Savage, S. (2013b). Predicting social influence with faction sizes and relative status. *Social Science Research, 42*, 1346–1356.
Melamed, D., & Savage, S. V. (2016). Status, faction sizes, and social influence: Testing the theoretical mechanism. *American Journal of Sociology, 122*(1), 201–232. https://doi.org/10.1086/686943
Morgan, L. H. (1997). *Systems of consanguinity and affinity of the human family.* University of Nebraska Press. (Original work published 1871).
Newcomb, T. M. (1953). An approach to the study of communicative acts. *Psychological Review, 60*(6), 393–404. https://doi.org/10.1037/h0063098
Olson, M., Jr. (1965). *The logic of collective action: Public goods and the theory of groups.* Harvard University Press.
Popper, K. (1959). *The logic of scientific discovery.* Harper & Row. (Original work published 1934).
Poundstone, W. (1993). *Prisoner's dilemma.* Anchor.
Radcliff-Brown, A. R. (2018). *The Andaman islanders.* Franklin Classics. (Original work published 1922).
Skvoretz, J., & Willer, D. (1993). Exclusion and power: A test of four theories of power in exchange networks. *American Sociological Review, 58*, 801–818.
Udehn, L. (2001). *Methodological individualism: Background, history and meaning.* Routledge.
Von Neumann, J., & Morgenstern, O. (1944). *Theory of games and economic behavior* (Sixtieth-Anniversary ed.). Princeton University Press.
Wasserman, S., & Faust, K. (1994). *Social network analysis: Methods and application.* Cambridge University Press.
Weber, M. (1968). *Economy and society.* University of California Press. (Original work published 1918).
Willer, D. (1986). Vulnerability and the location of power positions. *American Journal of Sociology, 92*(2), 441–48
Yamaguchi, T., & Cook, K. (1990). Power relations in exchange networks: A comment on 'network exchange theory.' *American Sociological Review*, 55, 297–300.

CHAPTER 7

Elementary Theory

Today social researchers face this conundrum. With few exceptions, graduate programs world-wide teach much the same inductive research methods. Those methods support inductions from samples to populations. That is to say, they support induction from a few to the many. But inducing from a few to the many is not induction from particular to general knowledge. How to produce fully general knowledge is not a part of the program of teaching sociology. Because it is not, sociology cannot explain historical evets or predict future events. The only way to produce fully general knowledge that can explain and predict is through theory. This chapter shows how to build theory. The best way to show how to build theory is by employing an example. The example theory of this chapter is Elementary Theory.

Following Einstein, Elementary Theory is so named because it is composed of irreducible *elements* as simple and few in number as possible. The logical structure of this chapter is the structure of Elementary Theory. It begins with simple ideas and combines them to build increasingly complex formulations. Social actions are used to compose social actors and social actors to compose social relations. Then social relations are composed into a variety of social structures. Diagrams are built at each step of this process.

Central to Elementary Theory is to explain social power and its exercise. The definition of power we employ combines meanings of domination and exploitation from Weber and Marx, respectively (Willer, 1999).

D. Willer, P. Emanuelson, *The Logic of Theoretical Sociology*,
https://doi.org/10.1007/978-3-031-91709-7_7

By domination we mean, as Weber did, the control of one actor over another. By exploitation we mean, as Marx did, that one actor benefits at the expense of another. Frequently the two occur together. An actor is dominated by another in order to be exploited.

The objective of this chapter is to walk the reader step-by-step through the components of a social scientific theory showing how they are used together to build models that are employed to test theory. How experiments are designed to test theory we take up in Chap. 8 to follow. How theory is used to explain events outside the laboratory is taken up in Chap. 9. Those of our readers acquainted with more successful sciences will recognize that we freely borrow theoretical methods from other sciences and adapt them to the needs of social science. Importantly, the authors of this book do not claim that the methods we now introduce are the only way to build social theory. Instead, we do claim that formal social theories can be built using the same methods as employed by successful sciences.

Social Actors

Social actors decide and act. There are two and only two kinds of actions which social actors can take. As displayed in Fig. 7.1, those two are positive and negative sanctions. A sanction is positive if the receiving actor prefers to receive it than not. A sanction is negative if the receiving actor prefers not to receive it. Everything else in the theory bears upon which type of sanction flows and the size of its effect.

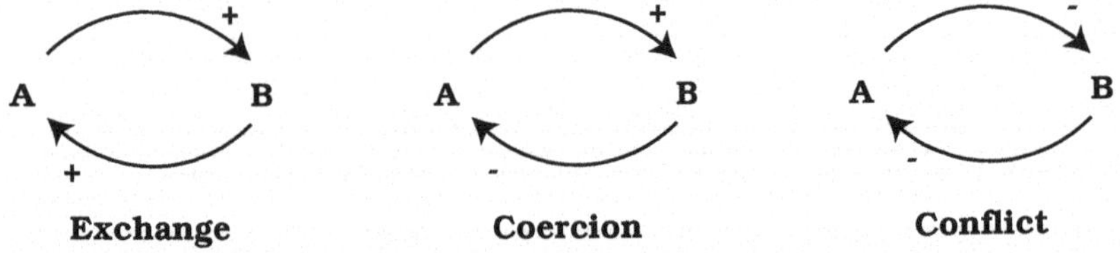

Fig. 7.1 Models of sanctions and relations

The receiving actor, 'B' in the figure, does not decide whether or not the sanction is sent. Only the sending actor, 'A' in the figure, decides whether the sanction is sent. As in the figure, sanctions always have signs of reception, signs that indicate their positive or negative effect. As shown in later diagrams, they may also have (1) signs of transmission indicating the effect on the transmitter of sending the sanction, and (2) numerical values indicating the amount of effect of the sanction on either the transmitter, the receiver or both (Emanuelson & Willer, 2018a, 2018b). Elementary Theory's first principle is:

P1: ALL SOCIAL ACTORS ACT TO MAXIMIZE THEIR EXPECTED PREFERENCE STATE ALTERATION. (Willer, 1999, p. 30)

Acting on the first principle, actors select relations and structures in which to act. How they act there requires something more than P1 to predict.

Social action is what social actors do. An individual choosing among fixed alternatives, as in comparing food prices in a grocery store, is not acting socially. That decision situation is called 'parametric' (Elster, 1986, p. 7). Action that takes into account the actions of others is social action. For Weber social action is the point of departure of sociology (1918/1968) as it is for the theory of games of von Neumann and Morgenstern (1944, p. 11). The decision situation in which "each actor's best choice of action depends upon the action he expects the other to take" is 'strategic' (Schelling, 1970, p. 86). In Elementary Theory social actors, when relating to other actors, always decide strategically.

Combining the two sanctions gives the three pure types of social relations: exchange, coercion and conflict. These are displayed in Fig. 7.1 (Willer, 1987, p. 63). The interdependences inherent in each of the three relations are the initial conditions from which flow negotiations over exchange ratios, rates of coercion and conflict (Willer, 1999, pp. 40–46). While differing in content, negotiations occur in all three types. When negotiations successfully link actions, the two actors are in agreement. When not, the two are in confrontation.

Figure 7.2 shows possible actions for actors in each of the three relations. For example, Fig. 7.2a shows the actions arrayed in order of preference for each actor in the exchange. Figure 7.2a shows that A prefers receiving B's sanction to reciprocal exchange, which in turn is preferable to no exchange. The least favored outcome is sending A's sanction while

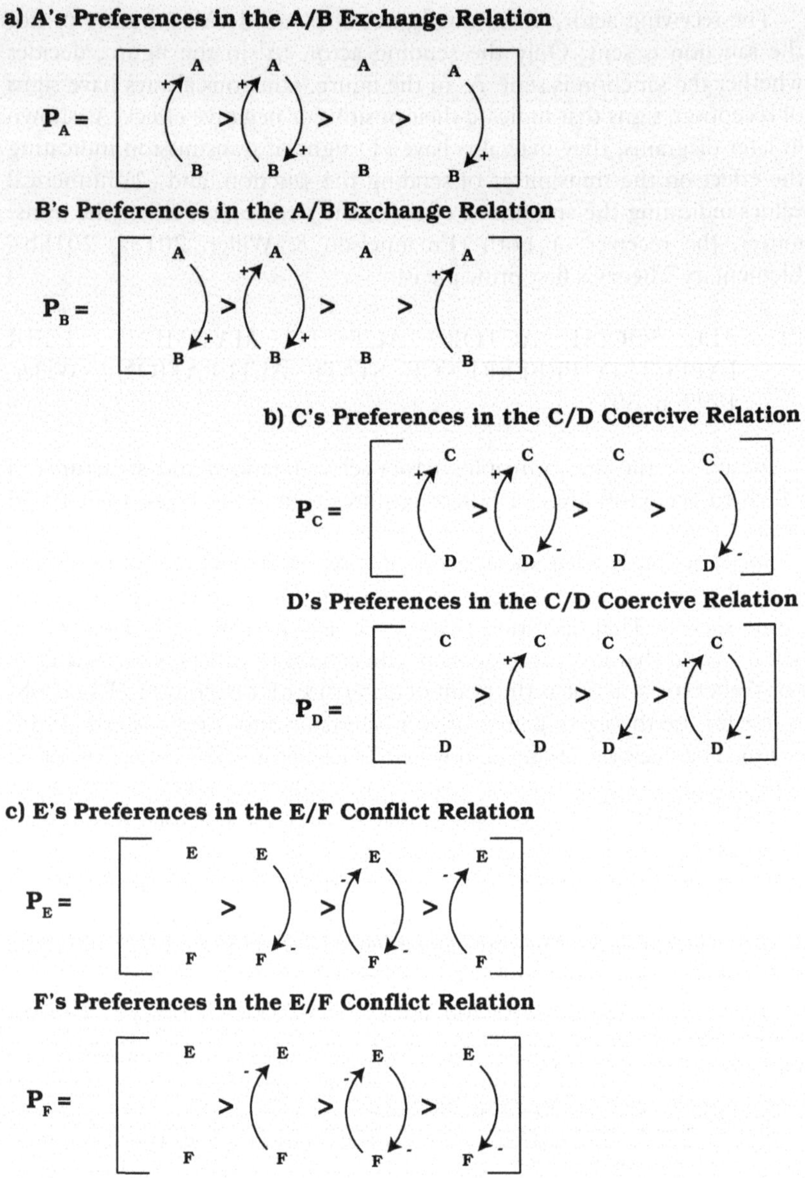

Fig. 7.2 Preferences for outcome by actor and relation type

not receiving B's. B's preferences are the mirror image of A's. Comparing preferences for coercion (7b) and conflict relations (7c) also shows mirror images of A and B. At issue is how do actors resolve these orderings such that they can act jointly (Willer, 1999, pp. 32–37).

This diagram shows why social actors negotiate, and how negotiations lead away from confrontation and toward agreement. B initially holds a single sanction worth 10 units to A and nothing to B while A initially holds 9 units each worth one to A and to B. If exchange occurs B will send its single sanction and, in exchange, A will send 'x' sanctions to B. Here are the payoffs for A and B.

	x = 1	2	3	4	5	6	7	8	9
P_A : Payoff to A :	9	8	7	6	5	4	3	2	1
P_B : Payoff to B :	1	2	3	4	5	6	7	8	9

An optimizing A would send only one sanction to B in exchange for A's single sanction worth 10 for a maximal payoff, $P_A = 10 - 1 = 9$. An optimizing B would seek to gain maximally 9 of A's sanctions in exchange for his one sanction worth 10 for a payoff of $P_B = 9 - 0 = 9$. Optimizing actors begin and end with maximally opposed offers, and since they do, it seems that exchanges cannot occur. But exchanges do occur. Let's look again at A's and B's interest situations.

Recognizing that failure to exchange is *not* costless, it is possible to define the two interests on which each actor acts: the interest in gaining the maximal payoff and the interest in avoiding the confrontation payoff. The first actor's interest in gaining the maximal payoff is $P_A\text{max} - P_A$: the difference between the maximal payoff, $P_A\text{max}$ and the payoff, P_A; and similarly for B. The interest in avoiding confrontation is $P_A - P_A\text{con}$: which is the difference between the payoff P_A and payoff at confrontation—and similarly for B. Then A's resistance is:

$$R_A = \frac{P_{A\max} - P_A}{P_A - P_A\text{con}}$$

and similarly for B.

Actors in Relationships

Figure 7.1 displays the two sanction types, positive and negative, and also shows the three pure types of social relations which are each composed of two sanctions (Willer, 1999, pp. 24, 26). Sanctions may also affect the preference state of the actor sending the act. Figure 7.3 displays two bisigned digraphs that show the two forms of exchange (Willer, 1999, p. 29). As indicated by the Figure, in economic exchange the signs indicate transactions that are losses to the sender and gains to the receiver. Here interests are opposed but complementary. In social exchange both sending and receiving are gains. The two types are associated with social distance. Economic exchange could represent strangers buying and selling while social exchange could represent friends exchanging favors.

The nodes of the Figs. 7.1 and 7.3 are social actors whose decisions and actions render the diagrams dynamic. The actors are endowed with values and beliefs upon which they decide and act. Borrowing from Marx, the authors prefer to employ 'reflective' actors whose values and beliefs accurately reflect the social relations and structures in which they act, as well as the values and beliefs of other actors acting in the structure. In other words, Elementary Theory's reflective social actor has a consciousness that accurately reflects its own meaning and meanings of other actors. Structures with reflective actors are the simplest to build because, once the structure is drawn, the values and beliefs of all actors are given. Reflection is not a principle of the theory. It is an assumption that can be withdrawn when actors have false consciousness—which is to say inaccurate values and/or beliefs. Legitimacy is frequently a result of inaccurate beliefs.

Bargaining and Resistance

Consider the interest situation of any actor 'A' in an economic exchange relation shown in Fig. 7.4. A quantity of ten units is at issue in the relation. $P_A = 9$ is A's best outcome. That is to say $P_A = P_A max = 9$. But that payoff

Fig. 7.3 Two forms of exchange relations

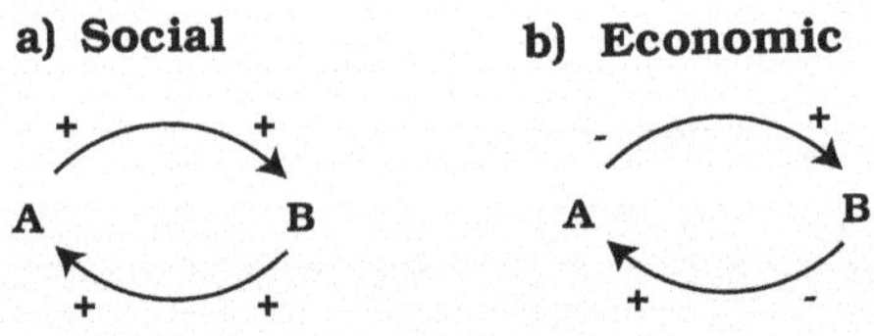

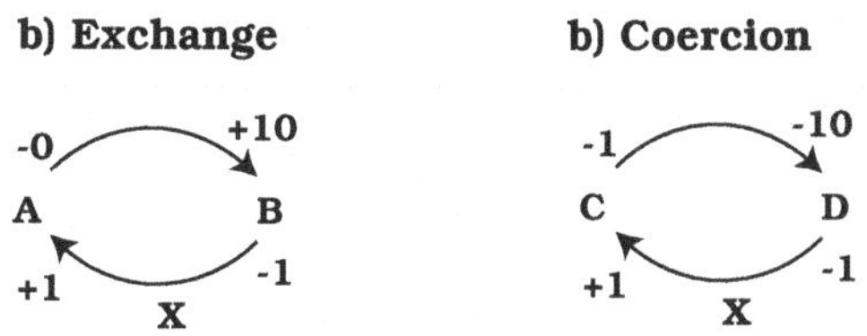

Fig. 7.4 An exchange and a coercive relations each with transmitting and receiving signs and sanction values

is most risky to A because B would only lose one for failing to agree. That is to say, $P_B - P_B\text{con} = 1 - 0 = 1$. Similarly, $P_B\text{max} = 9$ is most risky to B because $P_A - P_A\text{con} = 1 - 0 = 1$. Bargaining is a process of balancing payoff to self by increasing the risk of other. For this simple exchange relation, the only way self can increase risk for the other is to accept a lower payoff. The point at which payoff and risk balance for both self and other is at equal resistance. Therefore, agreement occurs at equiresistance when the interests in a better payoff $P_A\text{max} - P_A$ and $P_B\text{max} - P_B$ and risk $P_B - P_B\text{con}$ and $P_A - P_A\text{con}$ are balanced. The second law is the equation for resistance (Willer, 1999, pp. 43–44):

$$R_A = \frac{PA\text{max} - PA}{PA - PA\text{con}} = \frac{PB\text{max} - PB}{PB - PB\text{con}} = R_B$$

And the second principle is:

PRINCIPLE 2: FOR REFLECTIVE ACTORS, AGREEMENT OCCURS AT EQUIRESISTANCE.

When deciding actors interact with other deciding actors, Principle 1 does not give a unique solution. Thus, the need for Principle 2 and the resistance equation. For the exchange relation given above,

$$R_A = \frac{9 - PA}{PA - 0} = \frac{9 - PB}{PB - 0} = R_B$$

And, by symmetry, bargaining results in compromise at $P_A = 5$ and $P_B = 5$.

Resistance is much more flexible and broader in scope than it may at first appear. For example, resistance also applies to the coercive relation. For a simple coercive relation, C the coercer holds a single negative sanction that costs $P_C = -1$ to transmit to the coercee, D and negative ten to

D to receive ($P_D = -10$). The coercee, D initially holds ten resources worth one to D and one to C. Then

$$R_C = \frac{9 - PC}{PC - (-1)} = \frac{0 - PD}{PD - (-10)} = R_D$$

P_D max = 0 because D's best outcome is to be left alone by C. At agreement $P_C = x$, and $P_D = -x$. Cross multiplying and solving, x = 4.5, $P_C = 4.5$ and $P_D = -4.5$. The compromise point is for D to send 4.5 resources to C. Relative to exchange, the equiresistance compromise point has shifted to favor only C, the coercer. Since only C benefits, equiresistance is not counted as equal power. Instead, C, the coercer is exercising power over D, the coercee.

Closely related to the exchange relation are *null connected* exchange structures, two of which are displayed in Fig. 7.5. In the first structure,

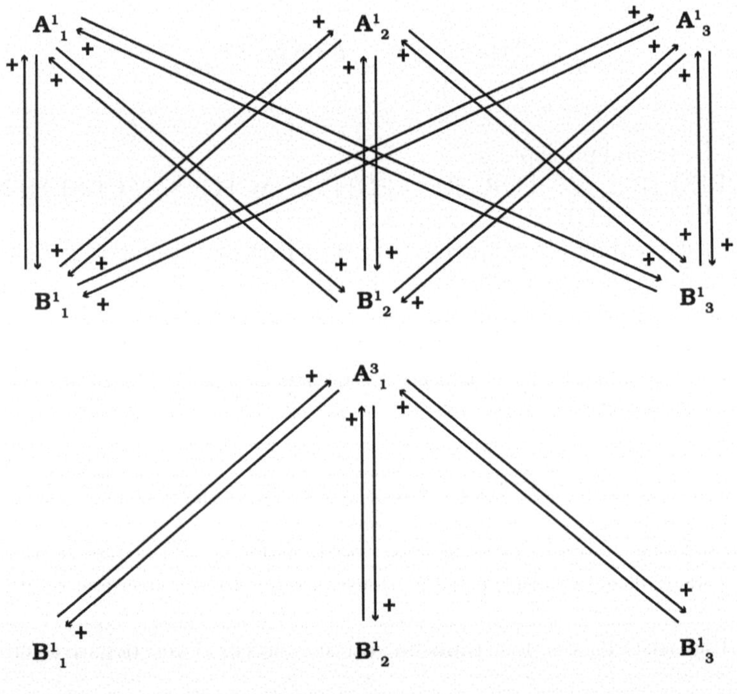

Fig. 7.5 Two null exchange structures

three As are exchanging with 3 Bs. Each actor's superscript gives the number of exchanges that actor may complete for each cycle of the structure. Note that all actors exchange once but for the centrally placed A actor in the centralized structure. That A exchanges three times. Each node initially holds exactly the same resources as in the exchange relation modeled above. No power condition is present that would change the prediction from that of a single exchange relation. The exchange ratio for 7.5 is exactly as it was in the exchange relation already modeled.

The centralized 7.5 structure is also null connected. Resources are held constant relative to the exchange relation. The central A has three resources each worth 10 to a B while each B initially holds ten resources valued by A and B. Again, there is no condition that would alter the exchange ratio from that of the single exchange relation. Therefore, the exchange ratio in each of A's three relations is the same as that of the exchange relation considered alone. Here we see that, contrary to the belief common across the social sciences, network centralization is not itself a power condition.

Structural Power I: Exclusion

Structural power is produced by the way relations are connected at a position. Exclusive connection is present when a position has a surplus of relations in which to exchange (Willer & Anderson, 1981; Willer, 1999; Skvoretz & Willer, 1993). For example, in Fig. 7.6, A is exclusively connected to two exchange relations identical to those introduced above. Being exclusively connected, A may exchange with B_1 or B_2 but not both. Let the interaction begin with B_1 making an offer to A to pay 2 for A's resource. (Remember that B's resources are lumpy.) Then B_2's best outcome has declined to $P_B max = 10 - 3 = 7$ and A's worst outcome, which is to accept the offer, is $P_A con = 2$. Plugging these values into resistance gives

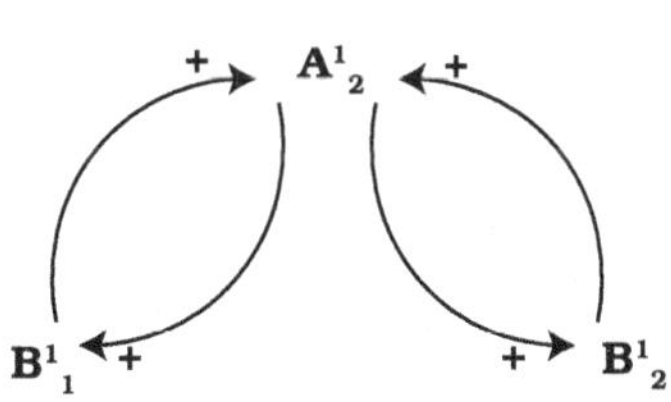

Fig. 7.6 A 2-branch exclusive exchange structure

$$R_A = \frac{9 - PA}{PA - 2} = \frac{7 - PB}{PB - 0} = R_{B2}$$

Since $P_A = x$ and $P_B = 10 - x$

$$R_A = \frac{9 - x}{x - 2} = \frac{7 - (10 - x)}{10 - x - 0} = R_{B2}$$

And solving, $x = 6$, $P_A = 6$ and $P_{B2} = 4$.

In words, B_1's offer of $x = 2$ has impacted A's and B_2's initial conditions. Now $P_A$con, A's worst outcome, is not zero but 2. And P_{B2} max, B's best outcome, is reduced from 9 to 7. Cross multiplying and solving for equiresistance, $P_A = 6$ and $P_B = 4$. Further steps will result in $P_A \rightarrow P_A$max = 9. A's payoff approaches A's maximum payoff. More generally, A's payoff will approximate the maximum possible for the exchange relation. Exclusion is a power condition. As will be seen, exclusion is the most efficacious of all power conditions thus far discovered.

Strong power structures that function like the one just analyzed include any network in which a set of nodes that cannot be excluded exchange only with a set where some nodes are always excluded. Resistance predicts that there will be a negotiation process that is not bargaining to compromise. Instead of bargaining, the negotiation process is iterative. The end point of the process is approached step-by-step by low power excludable positions making better and better offers to high power positions that never are excluded.

Resistance infers that offers are from low power to high power positions because, failure to make better offers results, as the reader can calculate, in zero resistance for those who are excludable. With that inference ET's analysis of power due to exclusion has now falsified two widely held beliefs about power. The first, as already seen, is that power is produced by the central placement of power positions. That belief appears to rest on a confusion of the effect of central placement with the effect of exclusion. The second falsified belief is that the exercise of power is characterized by active demands on the part of the powerful and passive acceptance of those demands by those low in power. It should be mentioned that these falsifications are due, not to resistance theory alone, but also to observation of the phenomenon of power in a series of experiments that included cross-national experimental investigations.

Marx discovered power due to exclusion that affects wages under capitalism. According to Marx,

> (T)he general movement of wages are exclusively regulated by the expansion and contraction of the industrial reserve army...by the varying proportions in which the working class is divided into active and reserve army. (1867/1967, p. 596)

In Marx's example the non-excludable positions were the members of the capitalist-class and the excludable positions were members of the working-class. There is no doubt that the effects of exclusion are by no means limited to capitalist labor markets—which is to say limited to time and place.

There undoubtedly are many other structures in which power is due to exclusion but that appear quite different from the exclusionary structures thus far considered. For example, there is no doubt that power in bureaucratic hierarchies is due to the presence of exclusion between each pair of levels. The design of bureaucracies, as Weber noted, includes systems of promotion. The official, acting as superior, evaluates each subordinate, an evaluation that translates into chances of promotion and salary increments for the subordinate. The official, acting as subordinate, is subject to evaluations that are a part of the same system of promotion. It follows that all officials but those at the bottom and the ultimate head act as both superior to those below them and subordinate to those above. This system of promotion builds exclusion into the structure between each pair of positions.

Structural Power II: Inclusion

A position is inclusively connected when it must complete agreements with *all* connected relations to gain payoffs from any. Beyond the theoretically pure case just defined, inclusion can be mixed with exclusive connection, with null connection, or with both (Szmatka & Willer, 1995; Willer, 1999; Simpson & Willer, 2005). Let's first take up the pure case of inclusive connection to show how resistance is applied to predict exchange ratios and infer interaction processes. Then take up mixed connection types. This analysis concludes by introducing ordering, a variant of inclusion.

The simplest case of inclusion occurs in Fig. 7.7a where two positions, B and C, are inclusively connected to a third, A. When exchanges occur sequentially, inclusion does not affect the first exchange. Therefore, the exchange ratio is as it was with null connection, an equal split at 5 – 5.

Here are the initial conditions for the second exchange:

$$P_A \max = 9 \quad P_c \max = 9$$
$$P_A \text{con} = -5 \quad P_c \text{con} = 0$$

Due to inclusion, the payoff to the central A position from the first exchange will be lost if it fails to exchange in the second. Therefore $P_A$con = −5. Resistance is

$$R_A = \frac{9 - PA}{PA - (-5)} = \frac{9 - PC}{PC - 0} = RC$$

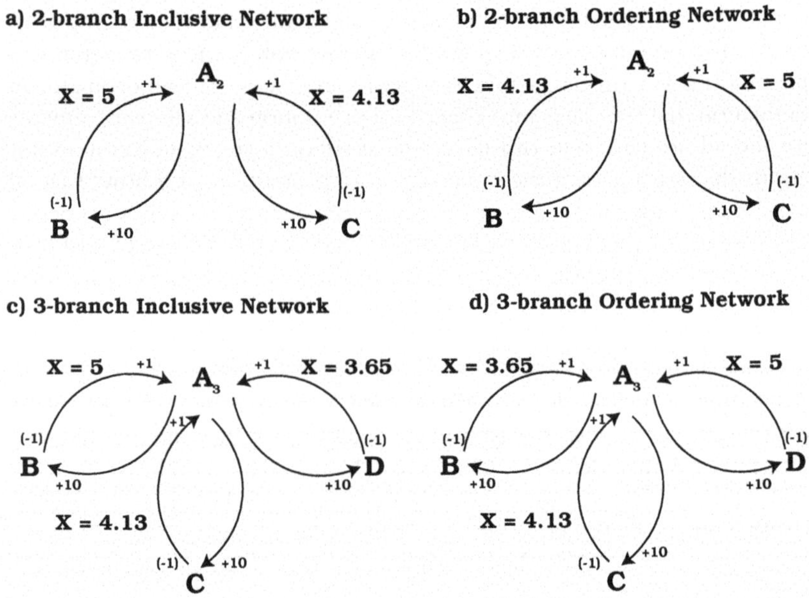

Fig. 7.7 Sequential exchanges for ordering and inclusion in a 2-branch network

Cross-multiplying and solving, $P_A = 4.13$ and $P_C = 5.87$. Now add another exchange relation, A – D and inclusively connect it to A. In negotiating with D, if A fails to exchange, it would lose payoffs from both earlier exchange: $P_C con = 5 + 4.13 = 9.13$. Plug those values in as above, for the third exchange $P_A = 3.65$ and $P_D = 6.35$. More generally, when exchanges at an inclusively connected position occur in sequence, equiresistance and exchange ratios shift, step by step, increasingly favoring the peripheral nodes.

However, exchanges need not occur in sequence. In a simple inclusively connected network, the peripheral actors know, as we do, that exchanging last gains the best exchange ratio. Therefore, unlike exclusion where the power process is driven by peripheral actors seeking to exchange first, in inclusion peripheral actors seek to exchange last. The struggle to exchange last pushes exchanges closer and closer to simultaneity. For simultaneous inclusion, let *Q* be the number of inclusively connected exchanges. For any exchange negotiation, there are Q – 1 other exchange relations that can affect it. Therefore $P_A con = (Q - 1)\ P_A$ and

$$R_A = \frac{PA\max - PA}{PA - \left(-[Q-1]\right)PA}$$

and simplifying

$$R_A = \frac{PA\max - PA}{QPA}$$

In general, the effect of inclusion, while increasing with Q, is always much smaller than the effect of exclusion. This difference is due to inclusion affecting only *one* part of the resistance factor, Pcon of the inclusively connected position. By contrast, exclusion affects both resistance factors of the exchanging actors. Both the Pcon of the exclusively connected position and Pmax of the peripheral positions are affected. At this point it should be mentioned that, like exclusion, in inclusionary structures the lower power actors are more active than higher power actors. Here, however, the 'activity' is not to exchange first, but to exchange last.

Ordering is a variant of inclusion. It occurs across relations when a node must exchange in a particular order (Corra & Willer, 2002). Let a central A be constrained to exchange with B, C and D in that order. As displayed in Fig. 7.7b, the effect of ordering is the reverse of inclusion. A's exchange with B is affected by the chance that A will fail to exchange with C. Thus, looking to Fig. 7.7d P_{A}con is largest in the negative direction in A's first exchange, but smaller in A's second exchange and has no effect on A's third exchange—and similarly for further ordered relations. Positions may be constrained by both inclusion and ordering. As Corra (2008) has shown, when both inclusion and ordering are present, their effects are additive.

Structural Power III: Some Mixed Cases

When inclusion and/or ordering occurs with exclusion, the joint effect is that exclusion overwhelms the effects of inclusion, ordering or both (Szmatka & Willer, 1995; Simpson & Willer, 2002). As will be shown, inclusion and ordering are so fragile that the effects of either or both can be eliminated by null connection.

There is no literature on inclusive connection or ordering, not because the effect of either is so very weak. It isn't. Instead, it has not been recognized because it is so easily eliminated by exclusion and markets are full of exclusion. As the reader can calculate, the order in which exchanges are negotiated and occur can determine which type of connection dominates. For example, an exclusively–inclusively connected actor, who, over a series of exchanges, recklessly uses up all its exclusions in early negotiations, may find itself inclusively connected over its final exchanges. Alternatively, successful collective action by peripherals can contain the effect of exclusion and use inclusion to gain favorable exchanges.

Nevertheless, some of the effects of inclusion have been recognized, if dimly. For example, organizations such as automobile manufacturing corporations that assemble parts into marketable wholes will find themselves in inclusively connected supplier networks. Cars without pistons, gears or many other components cannot be sold. The organizational literature, though not recognizing inclusion and its effects, has commented extensively on how to deal with potentially powerful suppliers. Henry Ford was said to be paranoid for "integrating backwards" by buying his suppliers. At one point in time, Ford Motor Company owned its own mines for iron, Great Lakes ships to carry the ore to Detroit, forests to supply wood for station wagons and so forth. Far from paranoia, incorporating all those functions under a single ownership structure eliminated the effect of inclusion. In Toyota's supply system, stock comes to be shared between Toyota and each of its suppliers. Like Ford, Toyota's extension of ownership eliminates the effect of inclusion.

Students of logic will have noticed a parallel between structural power conditions and sentential connectives. The relation of the two should now be made explicit. Let's begin by defining N, M and Q.

N_i is the number of relations connected at the position *i*.

M_i is the maximum number of relations in which *i* can exchange.

Q_i is the minimum number of exchanges *i* must complete.

By definition, $N_i \geq Mi \geq Q_i \geq 1$ and *i* is singularly connected when $N_i = M_i = Q_i. = 1$.

Sentential Connective	
i is exclusively connected when $N_i > M_i \geq Q_i. = 1$.	Exclusive 'or'
i is null connected when, $N_i = M_i > Q_i. = 1$.	Inclusive 'or'
i is inclusively connected when, $N_i = M_i = Q_i. > 1$.	'and'

Coercion: Relational and Structural Power

Compared to power in exchange networks, coercive power has received little attention. That is a mistake and perhaps a result of ethnocentrism. In today's world, almost everyone lives in a capitalist society and, as a result, engages much more intensively in exchange than in any other kind of relation. That economic exchange would be quite impossible without the coercive enforcement of property rights seems to be known only to legal scholars and some few economists. A little further thought will lead to the understanding that, as the political states of Europe came to control and eventually monopolize the employment of negative sanctions, only exchange remained for day-to-day relations. Today we intensively engage in exchange because the alternative of privately exercised coercion is blocked by the modern state.

The focus on exchange has pernicious effects. For example, Blau (1989, pp. 116–117) treats withdrawal of "regularly supplied" positives as the exact equivalent of sending negatives, thereby collapsing coercion into exchange. To be consistent with that formulation, the antebellum slave plantation would have to be treated as simply another mid-sized economic enterprise. Contrary to Blau's reductionism, in the U.S. Civil War thousands died fighting for and against slavery. To be consistent with Blau's treatment of coercion, the American civil war had to have been based on a confusion. Conversely, that so little has been done to investigate coercion has this positive implication—that there is so very much to learn. The investigation of coercion is open to discovery.

The coercive social relationship occurs in a dyad when one actor C, the coercer, holds one or more negative sanctions, and the other actor D holds one or more positive sanctions. Here are the initial conditions for a prototype coercive relation that has been used in laboratory experiments.

C holds one negative sanction. For each transmission $P_C = -1$ $P_D = -10$
D holds 10 positive sanctions. For each transmission $P_C = 1$ $P_D = -1$

When D's sanctions are lumpy and each valued at 1, there are nine possible coercive rates ranging from D sending one sanction to C to D sending nine to C. All nine are preferred by D to receiving C's negative. Applying resistance, we note that D's best outcome is for no sanctions to be transmitted. So $P_D max = 0$.

$$R_C = \frac{9 - Pc}{Pc - (-1)} = \frac{o - PD}{PD - (-10)} = R_B$$

Cross-multiplying and solving, PC = 4.5 and PD = – 4.5. C, the coercer is exercising power over D, the coercee. Importantly, this power exercise is wholly produced by the coercive *relation*. Unlike the exchange relation where power is only produced structurally, the coercive relation itself produces power exercise.

However, more power can be produced by a coercive structure. Here is an example of a strong coercive power structure. Connect six coercive relations of the kind introduced above at C the coercer (see Fig. 7.8a). There is a rule of the model that the coercer must send at least one negative sanction to a coercee for each cycle of the structure. Coercees will compete to avoid receiving the negative sanction by sending to the coercer increasing numbers of positive sanctions. The end point of the process is that the coercer receives nine positives sanctions from five coercees while the sixth coercee receives C's negative sanction. Here the effect of the structure is to double the rate of coercion from 4.5 to 9. In Chap. 9, applications of the theory toolbox use the distinction between null and strong power coercive structures to differentiate feudal and slave coercive structures.

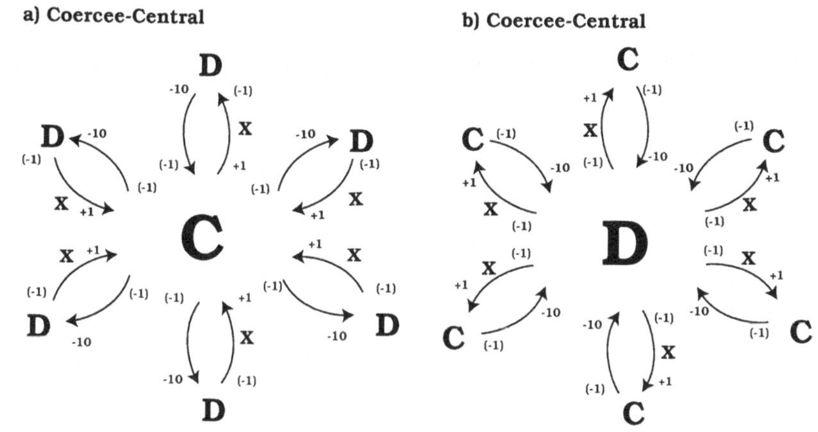

Fig. 7.8 Centralized coercive structures

Beyond increasing the rate of coercion, structures can also decrease that rate. Using again the coercive relation employed above, place the coercee in the central position and six coercers connected to it (see Fig. 7.8b). Make it a rule of the structure that the coercee can select which *one* of the coercers with which it will form an agreement. Again, a competitive process is expected, but in this case, the coercers are competing for the opportunity to coerce the coercee. The rate of coercion will approach zero. A frequently encountered example of this structure is in conjunction with mobile capital. States compete to attract mobile capital, reducing taxes to zero or below. Beyond the modeled structure developed here, coercees will pay capital to settle in their domain of coercion. Weber recognized the effect of mobile capital on coercion (1918/1968, p. 352).

Strategic Analysis

Strategic Analysis examines issues at the intersection of social structure and collective action (Willer & Skvoretz, 1997). The example analysis given here finds the game played by high power and the game played by low power actors. To find those games, the first step is to design a parametric strategy. This is that strategy: accept the best offer received or, if offers are tied, select randomly.

Fig. 7.9 A strong power exclusionary exchange structure

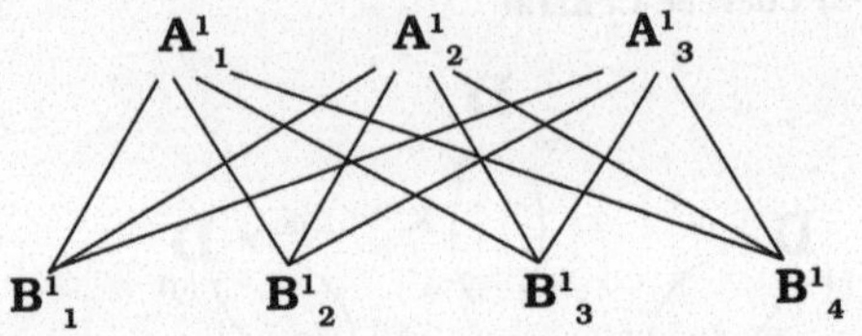

In the Fig. 7.9 structure, each A position is connected to all B positions and to no others while the Bs are each connected only to all the As. Each position exchanges only once per cycle of the structure. Therefore, one B is always excluded, and no A is necessarily excluded. Let each exchange relation contain a payoff of 24 points to be divided between the A and B exchanging partners. The analysis, as shown in Fig. 7.10, begins at the midpoint of the negotiation set where payoffs are equally divided 12 – 12.

Begin by attributing the parametric strategy to the As and derive the game played by the Bs. If all Bs cooperate offering 12 to the As, three of four will gain 12 and the fourth B is excluded for an expected payoff of 12 × ¾ = 9 which is the *R* payoff. If one, two or three Bs defect, they evach receive the *T* payoff of 11 while the fourth B is excluded for a payoff of 0, the *S* payoff. If all Bs defect, the payoff is 11 × ¾ = 8.25, the *P* payoff. Since $T > R > P > S$, the Bs are initially playing the Prisoner's dilemma (P/D) game where defection is the dominant strategy. When the Bs defect, A – B exchange payoffs become increasingly unequal favoring the As until 3 – 21 where exchange stops because the game is no longer P/D.

Attributing the parametric strategy to the Bs, each accepts its best offer or selects randomly. The result is that the As play the Privilege Game. The cooperation or defection of any A is entirely independent from the decision of the other As. Therefore, the dynamics of the Fig. 7.9 structure is determined by the P/D game played by the Bs. Because the Bs defect, the As exercise nearly maximum power over the Bs.

The foregoing is explained in greater detail in Borch and Willer (2006) together with a report of an experimental test. The result of the experiment was a good fit with the predictions of the Strategic Analysis. Furthermore, allowing A experimental subjects to act collectively had absolutely no effect on the payoffs of the structure, whereas allowing the B's to act collectively resulted in the Bs countervailing power such that 12 – 12 exchanges resulted.

t_0 = NPD

Number of Other Defectors	C	D
0	9	11
1	8	11
2	6	11
3	0	8.25

t_1 = NPD

Number of Other Defectors	C	D
0	8.25	10
1	7.33	10
2	5.5	10
3	0	7.5

⋮

t_8 = NPD

Number of Other Defectors	C	D
0	3	3
1	2.67	3
2	2	3
3	0	2.25

$t_9 = \neg$ NPD

Number of Other Defectors	C	D
0	2.25	2
1	2	2
2	1.5	2
3	0	1.5

Fig. 7.10 A defection chain

There are significant implications of the Strategic Analysis of the Fig. 7.9 structure for macrostructures. Economists view capitalist–worker relations as occurring in a labor market in which the two are engaged in exchanges that are in some sense equal. From the point of view of the research just reported, nothing could be further from the truth. Given a non-zero unemployment rate, capitalists are advantaged because they are playing the Privileged Game while the workers are disadvantaged being subject to the dilemma of the P/D game. We are not surprised that explanatory theory has social value implications. Economic theory's understanding of social structures is, however, restricted to markets of a particular kind. Strategic Analysis is not so limited.

Conclusion

Beginning with its simplest elements, positive and negative sanctions, Elementary Theory is built, step-by-step, to trace the effects of structure on the quantities of sanction flows, and thus power exercised. The result is a quite general theory of power in social structures.

Building the theory's formalism was furthered by its orienting perspective, the social theories of Marx and Weber. For example, the smallest unit in Marx's theory is the actor's consciousness while Weber's smallest element is the actor's meaning. But consciousness and meaning are not fundamental elements. Both can be cut into three elements that *are* fundamental. Those elements are (1) the actor's values, (2) the actor's beliefs and (3) the mechanism for actor's decisions to act.

Here is a potential problem. Actor A's beliefs about actor B include B's beliefs about A, B's beliefs about A's beliefs about B and similarly in infinite regress. Simplification is needed. That simplification is 'reflection'. Inspired by Marx, each actor's beliefs are a reflection of the social structure in which the actor acts. Given that simplification, a decision mechanism can be deployed, resistance calculated, and power exercise predicted.

In Elementary Theory reflection is not a law nor is it a mechanism. It is a simplifying assumption and, as such, can be withdrawn as needed. It may be useful to think of 'reflection' by analogy to friction. We gain simplicity by using reflection much as Galileo and Newton gained simplicity by ignoring friction. Note that the exercise of influence violates reflection because beliefs are manipulated. Much of that manipulation may well be inculcating false beliefs.

Elementary Theory's use of diagrams to model social structures offers a number of advantages over theories that do not picture their phenomena. As already mentioned, the step-by-step introduction of the mechanisms of the theory are paralleled by, and governed by, the step-by-step introduction of elements used to diagram structure. Since the theory is built from simple to complex, it contains a logically consistent technical vocabulary. Furthermore, as elements are combined and recombined, an increasingly wide array of structures can be built and investigated. In this chapter the focus has been on relatively simple structures. However, the groundwork has been done for modeling more complex structures, including ones where sanctions flow through positions to affect outcomes many steps away in the structure (Willer, 2003).

The last issue taken up in this chapter was at the intersection of social structure and collective action. Working at that intersection, Strategic Analysis opened an array of new questions for theory to investigate. The next chapter, Chap. 8, shows how theory designs the experiments that test it. Then Chap. 9 shows how theory is applied in scientific explanation.

References

Blau, P. M. (1989). *Exchange and power in social life*. Transaction Publishers. (Original work published 1964).

Borch, C., & Willer, D. (2006). Power, embedded games, and coalition formation. *Journal of Mathematical Sociology, 30*, 77–111.

Corra, M. (2008). Inclusion and ordering: The compounding effects of two distinct but related structural power conditions. *Social Behavior and Personality: An International Journal, 36*(9), 1161–1178.

Corra, M., & Willer, D. (2002). The gatekeeper. *Sociological Theory, 20*, 180–205.

Elster, J. (1986). *Rational choice*. New York University Press.

Emanuelson, P., & Willer, D. (2018a). Elementary theory. In P. Burke (Ed.), *Contemporary social psychological theories* (2nd ed., pp. 194–224). Stanford University Press.

Emanuelson, P., & Willer, D. (2018b). Social structures in transition: Application of two theories to chiefdoms. *Socius: Sociological Research for a Dynamic World, 4*, 1–14. https://doi.org/10.1177/2378023118771758

Marx, K. (1967). *Capital*. International Publishers. (Original work published 1867).

Schelling, T. (1970). *The strategy of conflict*. Harvard University Press.

Simpson, B., & Willer, D. (2002). Applications of Elementary Theory to the social structures of antiquity. In S. C. Chew & J. D. Knottnerus (Eds.), *Structure, Culture and History: Recent Issues in Social Theory* (pp. 231–254). Rowman and Littlefield.

Simpson, B., & Willer, D. (2005). The embeddedness of collective goods. *Sociological Theory, 23*, 386–407.

Skvoretz, J., & Willer, D. (1993). Exclusion and power: A test of four theories of power in exchange networks. *American Sociological Review, 58*, 801–818.

Szmatka, J., & Willer, D. (1995). Exclusion, inclusion and compound connection in exchange networks. *Social Psychology Quarterly, 58*, 123–131.

Von Neumann, J., & Morgenstern, O. (1944). *Theory of games and economic behavior*. Princeton University Press.

Weber, M. (1968). *Economy and society*. University of California Press. (Original work published 1918).

Willer, D. (1987). *The and the experimental investigation of social structures.* Gordon & Breach.

Willer, D. (1999). *Network exchange theory.* Praeger.

Willer, D. (2003). Power-at-a-distance. *Social Forces, 81*, 1295–1334.

Willer, D., & Anderson, B. (Eds.). (1981). *Networks, exchange and coercion.* Elsevier/Greenwood.

Willer, D., & Skvoretz, J. (1997). Games and structure. *Rationality and Society, 9*, 5–35.

CHAPTER 8

Theory and Experiment

This chapter reviews theory-driven experiments to uncover their structure. Oddly, methods texts present Fisher design as the only way that experiments can be designed. Using that design, the fledgling student is to discover regularities, not the perfect regularities that connect causes and effects, but a likelihood of regularity. Random assignment to experimental and control groupings addresses Mill's problem of matching, but for a single difference while providing the basis for statistical testing. The use of random assignment does not produce groupings that are identical but for one difference. But it does produce groupings that are on average similar and that has been deemed sufficient.

Fisher claimed that the methodology he introduced was inductive. That claim is false. As Karl Popper (1934/1959) demonstrated in *The Logic of Scientific Discovery*, published a year prior to the publication of Fisher's (1935) *Design of Experiments*, there is no inductive method. If there were an inductive method, it would link the particulars observed in the lab to general knowledge about the world—rather like the well-known link between a randomly drawn sample and its population. But no such method has ever been proposed. Since there is no inducive method, there is nothing in Fisher's design that justifies inferences from laboratory to the world outside the lab. Had Fisher solved the induction problem, there would be no replication problem in the social sciences. But Fisher offered no solution to the problem of induction and the replication problem is

D. Willer, P. Emanuelson, *The Logic of Theoretical Sociology*,
https://doi.org/10.1007/978-3-031-91709-7_8

one result. While the best practice currently in sociology is not Fisher design, one example is offered to show how Fisher designs are devised.

While not inductive, contrary to Karl Popper, theories are also not deductively applied. Neither the theory itself nor any of its parts acts as a major premise nor does any part of the theory-designed experiment appear as a minor premise. Theory application is not syllogistic.

Furthermore, contrary to methods texts, best practice in contemporary experimental design owes little or nothing to Fisher. It does not construct and then compare experimental and control groupings. Instead, some of the best contemporary designs test hypotheses. Frequently a series of tests are offered to differentiate the hypotheses being tested from previous work.

Theory-designed experiments in their fully developed form begin with the construction of a theoretical model. The theoretical model then acts as a template for the construction of the empirical replica. Theories are tested in the laboratory by building empirical replicas that correspond to—are similar to—models built in the theory. When all of the components for modeling are a part of the theory, building the experiment is quite straightforward. The last experimental design discussed in this chapter has theoretical models and corresponding experimental replicas.

All of the theory-driven sociological experiments reviewed here are exemplary in that they take up significant issues in theory and offer resolutions to them. Since none are at all like the Fisher-designs recommended in textbooks, their success might cause some confusion. Nevertheless, most discussed here fall short of being fully theory-driven. They fall short because they lack the modeling procedure needed to build experimental replicas. Without a modeling procedure, a modeled social structure or any other needed model simply cannot be composed. Without a model a replica cannot be built, and a fully theory-driven experiment cannot be designed and run. The reviewed experiments are exemplary because they resolve issues in theory. Their only shortcoming is their lack of models and corresponding replicas. The result, as will be seen, is that they do not produce cumulative knowledge.

Outside the laboratory where the empirical situation cannot be constructed to replicate a theoretical model, the process of application is reversed. Given observed events to be explained, the theorist searches for a theory and model corresponding to the observations. Frequently, the complexities of the world can be explained only by employing multiple theories. When multiply theories are employed, they are referred to collectively as a 'theory toolbox'. Examples of how theory is applied outside the lab are the topic of Chap. 9.

One Fisher Design and Four Theory-Driven Designs

Let's now review the methodological structure of a few experiments. Only the first reflects Fisher-design. It is reviewed in order to show how it differs from current theory-driven practice. The remaining experiments are exemplary because they address and resolve issues in sociological theory.

Analytic Thinking

The study by Gervais and Norenzayan (2012) published in *Science* proposed to test whether analytic thinking promotes religious disbelief. For their design there are two distinct kinds of thinking; 'System 1' intuitive thought and 'System 2' analytic processing. The first was predicted to promote belief in the supernational while the second to undermine that belief.

Five studies were instituted. Study 1 measured results for subjects on three analytic thinking tasks and compared those results with three religious belief scales. Here is one of the analytic thinking tasks.

If it takes 5 machines 5 minutes to make 5 widgets, how long would it take 100 machines to make 100 widgets? _____ minutes

The correct answer is 5. Giving the incorrect answer, 100 minutes, was found to be related to positive responses on three religiosity and belief scales. Here is a sample from each of three religiosity scales.

> My faith involves all of my life.
> I believe in God.
> Angels exist. (2012, p. 494)

Study 2 used a visual prime. Subjects were shown one of two pictures, either Rodin's Thinker or a discus thrower. The hypothesis was that the picture of Rodin's Thinker and not the picture of the discus thrower would prime analytic thinking. Statistical tests confirmed the hypothesis.

It is difficult to imagine a science to which this study would contribute. It is easy to believe that the study was intended to be a self-parody, though of course it was not. But compare the analytic thinking task quoted above to following parody from The Devil's Dictionary:

> Major Premise: Sixty men can do a piece of work sixty times as quickly as one man.
> Minor Premise: One man can dig a posthole in sixty seconds; therefore—
> Conclusion: Sixty men can dig a posthole in one second. (Ambrose Bierce)

The picture of the Thinker shown in Study 2 is not what the authors apparently believe it to be. Rodin's thinker is not thinking analytic thoughts. The thinker first appears atop Rodin's Gates of Hell where he is contemplating the fate of souls entering hell—clearly affirmation of religious belief.

An experiment that follows Fisher design faithfully, as does Gervais and Norenzayan, is no guarantee of its scientific worth. To escape the limits of the lab, designs must address theoretic issues. But Fisher-design offers no guide, and no criteria for the relevance of formulations selected. The study was published in *Science*, the leading general science journal published in the U.S. That it was indicates to us that the editors need to develop criteria of relevance for in this case, they have lost their way.

Social Identity and Cooperation in Social Dilemmas

This study by Simpson (2006) draws on two distinct theories, Game Theory and Social Identity Theory to explain cooperation in social dilemmas. It is important to note that this study is by no means the first to draw on those two theoretics. At issue in Simpson's paper is not simply the joint application of two theoretical traditions, but exactly how the two work together. This discussion has two parts. The first explains core ideas of the two theoretics. The second reviews how Simpson applies them jointly.

Social Identity Theory rejects the long-standing assumption of economics and some Rational Choice Theory in sociology that people are motivated only by self-interest. In Social Identity Theory that interest, called proself, is distinguished from prosocial where consideration is given to others. In social dilemmas, prosociality may affect the belief in whether others will cooperate, or it may affect expectations concerning how actors weigh others' outcomes. Which one is a central issue? Turning to Game Theory, Table 8.1 gives Prisoners' Dilemma (PD) payoffs for the 2-person game. The R payoffs are for mutual cooperation, while the P payoffs are for mutual defection T and S payoffs are for unilateral defection and cooperation respectively. The game is PD because the payoffs are ordered $T > R > P > S$.

Table 8.1 Prisoner Dilemma payoffs for a 2-person game

	Player B: Cooperate	*Player B: Defect*
Player A: Cooperate	2, 2 R, R	0, 3 S, T
Player A: Defect	3, 0 T, S	1, 1 P, P

These payoffs can be further analyzed into greed and fear components. As explained by Simpson, "Greed is the temptation to free ride on others' cooperation. (greed = T – R) Fear is the motivation to avoid being exploited by non-cooperative others. (Fear = P – S)" (2006, p. 446). For the experiment to avoid cuing subjects, cooperation and defection alternatives were called Alternative 1 and Alternative 2, respectively. To produce 'minimal' groups, subjects were asked to express preferences for one or another painter whose styles are sharply contrasting. The majority of preferences for one or the other defined the subject's ingroup. Subjects then made a number of cooperation or defection choices. For each choice alter was identified as ingroup or outgroup.

The most interesting experimental predictions and outcomes occurred when the PD game was played sequentially. There are two ways to play sequentially: ego choses first or choses second. When choosing first ego is affected only by the fear component. Therefore, the ingroup – outgroup distinction should have no effect. When choosing second and other cooperates, however, ego faces the greed component. Therefore, there should be ingroup vs. outgroup effects with greater cooperation with the ingroup. Experimental results support these predictions. Subjects choosing first cooperated at very nearly equal rates with alters who were members of the ego's ingroup and alters who were outgroup members. Subjects choosing second, however, cooperated at almost twice the rate with ingroup alters than outgroup alters.

This study represents a considerable advance. For prosocials, earlier views simply traced cooperation to ingroup membership and defection to acts toward the outgroup. That difference was said to result from a norm of generalized reciprocity. The effect of generalized reciprocity was qualified by the fact that high levels of cooperation can be sustained in repeated dilemmas. Thus, identity was expected to have no effect on repeated dilemmas. By contrast, theory deployed by Simpson and supported by experimental results says that social dilemmas are best solved when the motivation to defect is from greed, but not when it is from fear.

Simpson's study highlights the superiority of laboratory experimentation over other forms of investigation. It is difficult to imagine how the investigator would find naturally occurring events with the contrasting conditions required to test Simpson's hypotheses. To be more precise, it is not difficult to imagine—it is a practical impossibility. Theory testing demands the control available only in the lab. The great shortcoming of laboratory research is, we are told, artificiality. The claim here is that all that is outside the laboratory is real in contrast to the lab's unreality. Were this thought to be taken seriously we would have to set aside all laboratory research in sociology along with all laboratory research in all the sciences.

The Status Solution to the Collective Action Problem

The collective action problem, at least in its modern form, can be traced to Olson's (1965) objection to Bentley's (1908/1949) "group theory of politics". According to Bentley, in a pluralistic political environment such as the U.S., people form groups to pursue their common interests, a formation that is explained by the common interest being pursued. Olson objects, asserting that, while it is rational for people to pursue individual interests, no rational person will pursue common or collective interests. They will not because any benefits that resulted would be received by all in the collectivity whether they paid the costs of gaining those benefits or not. Benefiting but not paying the cost is called 'free riding'. Free riding can be suppressed by coercion, but coercing creates a "second order" free riding problem (Oliver, 1980). Who will pay the costs of coercing?

Following Berger et al. (1972), R. Willer defines status as an individual's relative standing in the group based on prestige, honor and deference, a definition that also echoes Weber (1918/1968). Status is a selective incentive that solves the collective action problem, according to R. Willer (2009a, 2009b) because people gain valued respect for making contributions that benefit the group. At issue is not an exchange of contributions for status, but an enhancement of status by demonstrating motivation to benefit the group through their actions. Status enhancement in turn leads to increased contributions to the group in the future. R. Willer is not proposing that the individual's collective actions enhance status which, in turn, increases the individual's collective actions which then further enhances status in a positive feedback loop. Instead, collective actions and status come to form an equilibrium where the two are in balance.

Four experiments test these formulations and differentiate them from alternative explanations. To those ends R. Willer (2009a) proposed a variety of experimental designs. Study 1 included four parts. In the first part subjects played a public goods game. Endowed with $5 each subject played a sequence of five rounds. In each round up to $1 could be contributed to the public. Each contribution was doubled and distributed equally to all. High contributors were predicted to gain high status. Before Phase 2 began, subjects were asked to rate assigned partners as "honorable", "prestigious" and "respected". Phase 2 of Study 1 measured influence using the contrast sensitivity task in which subjects viewing a checkerboard of shaded squares estimated whether more were white or more were black (Berger et al., 1977). Subjects made two judgements, the second and final judgement after viewing the choice of a simulated other. The simulated other disagreed in 20 of 25 viewings. Subjects were paired with previously high or low contributors. The number of times subjects switched their judgements measured influence exercised. Phase 3 used the Greed and Fear of Greed games to measure subject's cooperation and the Dictator game to measure subjects' generosity. Subjects playing the Dictator game were given an endowment, all some or none of which could be contributed to a (simulated) other.

The results of Study 1 supported the theory of the paper. High contributing members were both more group motivated and higher status. Status was related to influence and high contributors gained greater cooperation than did low contributors.

Studies 2, 3 and 4 were designed to test further hypotheses and evaluate alternative explanations. For Study 2, reciprocity could explain at least some of the results of Phase 1. For example, it could be that the generous giving of A is reciprocated by B evaluating that A's status as higher. To cut through that possible link, subjects did not act in the public goods game—they simply observed it. Not acting in the public goods game, subjects' acts could not be reciprocated. Study 3 asked whether others who contributed a higher proportion were more group oriented and of higher status. Study 4 asked whether high status gained from group motivation would lead to higher willingness to contribute in the future. All those hypotheses were supported.

Prior to this paper the received view saw collective action motivated either by material rewards or by status in that the costs of acting collectively were paid for by valued status enhancement. This paper saw a relation between status and collective action, but for very different reasons. R. Willer showed that contributions lead to enhanced status because they signal a group motivation in which a concern with the group's welfare is increased relative to their own. Another implication of R. Willer's theory found in another paper (R. Willer, 2009b) is that the status theory of collective action does not give rise to second or higher order collective action problems. It does not because substantial evidence shows that people update their view of others and do not consider that updating a cost.

Once again, we see an investigation that simply could not be carried out using non-experimental research. As in Simpson's study, it would be quite impossible to discover natural cases, the variations of which align with the propositions put forward by R. Willer. Nor is there a problem of "external validity" as is frequently claimed. A detailed discussion of the external validity problem is found in Lucas (2003). It is Suffice it to say here that, given there is no inductive method, well developed theory-driven experimental research has a better claim for "external validity" which is to say 'generality' than any survey research we have seen.

Incentivizing Low Status

For more than five decades, Status Characteristics Theory (SCT) has steadily increased its scope by applying theory and designing theory-driven experiments. This is an unparalleled record in the social sciences. During that time a central idea driving the research has been the relation between status and influence and the effect of the two, separately or together on an array of social phenomena including, as just seen, a new solution to the collective action problem. Pervading this work has been the understanding that status orders are deeply embedded in the everyday lives of people and that achieving increased status in the prestige order is an important incentive favoring group activity.

Not recognized before Ridgeway and Nakagawa (2017) was why those of low status would accept their depressed status and, indeed, accept a status quo that places them at or near the bottom. It would be fair to say that, prior to Ridgeway and Nakagawa, SCT contained a contradiction between status as a motivator and the pervasiveness of status hierarchies. How could prestige orders exist when the disadvantaged, who form a substantial portion of those in the order, had no reason to accept it? Reasons could be found by evoking another theory. For example,

Elementary Theory would suggest that power structures, not based on consensus, coerce those low in status to accept their place in the prestige order. By contrast, Ridgeway and Nakagawa's solution rests on a hitherto unrecognized motive produced by the prestige order that incentivizes those of low status to continue within the order.

Ridgeway and Nakagawa hypothesize that those of lower status who give deference to those of higher status receive "a general, modest, but meaningful positive incentive…the dignity of being respected in the group as reasonable" (2017, p. 149). They emphasize that, though this motive is substantially weaker than motives linked to enhanced status, it is nevertheless effective and real in its consequences. Three hypotheses are proposed. The hypotheses are concerned with a 'second order expectation'—a belief of self about what others believe about self. Specifically, group members who expect to perform less well will anticipate that others will view them with respect were they to defer to and not resist others. Furthermore, those of lower status, having received respect, will be more committed to the group.

Four experiments tested the hypotheses. In experiments 1, 2 and 3, subjects were recruited using Amazon Mechanical Turk and formed into three person groups, one person being the subject and two fictive. The task, named Meaning Insight, was a ten item test matching the meaning of words purporting to be from an "early language" with English words. As reported to the subject in experiment 1, fictive teammates score 6 or 7 of ten while the subject scored 3. In experiment 2 the subject was told they scored 6, thus moving the subject to the middle status. Scores were treated as specific status characteristics. After reporting the scores, subjects were asked whether they thought others wanted them to have status and influence, exercise control over others in an upcoming task and act as leaders. These questions measured second order expectations—of which others were believed to expect of self. In experiment 3, a vignette was substituted for meaning insight. Experiment 4 used a random sample recruited through "YouGov". Its vignette described three person groups with members of differing status. Subjects were asked to evaluate groups in which low status members were either deferential or resistant. All four experiments supported Ridgeway and Nakagawa's hypotheses concerning second order motives.

Beyond having plugged a longstanding hole in SCT, Ridgeway and Nakagawa lucidly demonstrated the importance of second order expectations. The key word here is "lucidly" for it is far too easy for writer, for reader or both to be caught in a second order labyrinth of "she believes that he believes that she…" Nevertheless, second and higher order expectations cannot be ignored as more and more sophisticated theories are built. Earlier in this book it was suggested that, for many purposes, an actor's beliefs could be treated as a straightforward reflection of the actor's social environment. For example, Elementary Theory has exclusively used reflective actors. Given the accuracy of its predictions, it may be that investigation of second and higher order expectations—which is to say beliefs—is not needed for a theory of power structures. In light of Ridgeway and Nakagawa's work, however, it may well be time to reconsider second and higher order expectations.

Status, Numbers and Influence

As mentioned above, Status Characteristics Theory (SCT) has an unparalleled history in the social sciences of more than five decades of theory growth and theory testing. Tens, even hundreds, of characteristics, such as educational attainment and beauty, have been evaluated for status effects and linked to influence processes. Despite this diversity and notwithstanding the employment of terms like "power and prestige order" that seem to imply a multiplicity of related people and positions, SCT has long been a dyadic theory. As Melamed and Savage (2016) explain, "SCT is silent with respect to how status processes may interact with numbers in groups larger than two" (p. 1086). As a result of their investigation, SCT now speaks to status processes beyond the dyad.

In their 2013 paper, Melamed and Savage develop and test a theory integrating the effects of status and placement in groups of four. Their "unified model of social influence" (p. 1088) gains that unity from the relation of status and number in jointly reducing uncertainty. The integrative role of uncertainty in their theory stands in contrast to its largely implicit place in previous SCT research. They postulate that people are "cognitive misers' (p. 1088) who look first at the distribution of opinions and, if there is consensus, do not look further for status effects. Therefore, the closer the distribution of opinions to a 50/50 split, the greater the effect of status. It follows that high status individuals are largely insulated from influence by others.

The following three hypotheses are tested.

1. Whether status is associated with resisting influence.
2. How faction size is related to resisting influence.
3. How faction size moderates the effect of status on influence.

The tests were carried out on groups of four where three were simulated. Initial conditions were the proportion of simulants' scores that agreed with the subject's and the placement of the subject relative to scores of others.

The scores stemmed from the contrast sensitivity test often used in SCT research. Subjects judged whether there were more white or more black squares on a checkerboard. Scores from those judgments were simulated and fed back to subjects. The experiments had two parts. Phase one established the relative status of subject and simulants by feeding back fictive "scores" of subject and simulants. In Phase two, the subject was shown the contrast sensitivity checkerboard twice, first for an initial judgment and second for a final judgment. Phase two measured influence by counting the number of times the subject stayed with the initial judgement. The "stay" response measured resistance to influence. The results showed that the effect of status on influence was moderated by the size of the faction supporting the subject's judgment.

Melamed and Savage noted the limits of their study, including the smallness of their groups of four and the relative isolation of the experimental subject from other (fictive) group members. Nonetheless, their study is a major step forward for SCT, not just in expanded scope, but in their employment of uncertainty to integrate status and number. Furthermore, it opens the way for future research extending status process beyond small groups.

In Chapter 9 the discussion of status-based chiefdoms extends the application of Melamed and Savage's theory to a large status-lineage structure. A status-lineage structure is a prestige order. It governs the distribution of status and influence in a chiefdom. Application of their theory to a prestige order as large as a chiefdom requires a modeling procedure that links placement of positions to their status and thus their influence. Fortunately, Hage and Harary (1996) offer a modeling procedure for status-linage chiefdoms. Its application together with Melamed and Savage's theory helps to explain how chiefdoms function as they solve their collective action problems.

Elementary Theory-Driven Experiments

Elementary Theory-driven experimentation extends the role of theory in experimental design beyond that of the other social theories discussed here. Using its modeling procedure, the Elementary Theory experimenter begins by building a theoretical model, which is a wholly abstract construct for the phenomenon to be investigated. In building the model, the actors' values are given by the transmission and reception values of the sanctions that are the parts of the social structure. Having built the structure with its sanction valuations, the experimenter/theorist next specifies the type of connection across sanctions at each position (i. e. exclusion, inclusion, ordering, etc.). Seeking parsimony by invoicing "reflection", the theorist infers actors' beliefs from the structure and the actor's place in it.

The result is a parsimonious theoretical model for the social structure to be investigated. Adding the assumption that actors know that their values and beliefs are accurate, the model will contain no influence processes. To introduce influence, following Melamed and Savage, we would add uncertainty of beliefs by limiting information.

The next step is to build an experimental replica corresponding to the theoretical model. This step is an exact parallel to that of an architect constructing a building by consulting a blueprint. The building is the (empirical) replica and the blueprint is the (abstract) model. The experimenter/theorist builds the experimental replica by consulting the abstract model. The initial conditions of the replica are the initial conditions of the model. If the experiment succeeds, the final conditions of the replica are the final conditions of the model.

Here is an example. Shown in Fig. 8.1a is the theoretical model for a simple centralized exchange structure with one central A position and three peripheral B positions. Not shown is the contingency limiting the central A position to two exchanges. That limitation determines that one B will be excluded each cycle of the network. Shown in Fig. 8.1b is the replica pictured as the layout of a laboratory. In the layout, the central A is face-to-face with the three B positions who are minimally isolated from each other by low walls. Not shown are initial resource distributions that are consistent with the model. They are as follows. The A initially holds two resources each worth one point if kept by A and ten points to the B who acquires it. Each B initially holds ten resources each worth one point to B and one to the A. For the experiment, the resources are represented

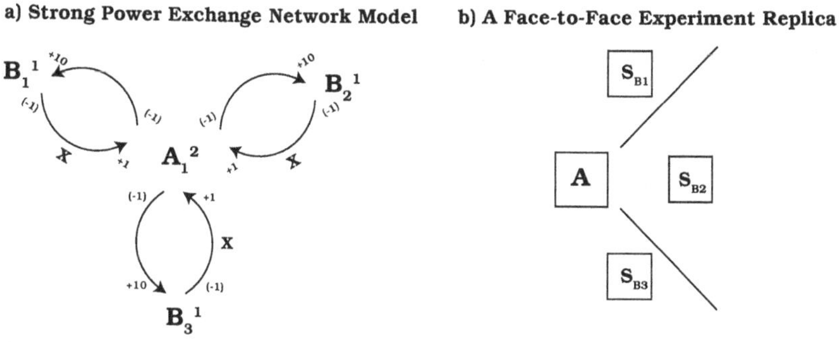

Fig. 8.1 A Br32 exchange network

by counters, perhaps colored poker chips that participants pass to each other at agreement. Cycles of exchange are timed such that all participants are given adequate time to negotiate and exchange. The number of cycles for each run is chosen, usually about ten, so that exchange ratios can attain equilibrium. The predicted equilibrium for this structure is for each B to send nine sanctions in exchange for one sanction worth 10 from the A. At the conclusion of the experiment participants are paid in money proportionate to the value of counters they have accumulated in exchange. Also not shown is the experimenter who times rounds, records the movement of resources, sums the totals for each participant, and pays off participates at the conclusion of the run. Alternatively, exactly the same design has been run using computers linked through a LAN.

Motivated by the payoffs from exchanging, subjects actively negotiate and exchange resources. The subjects' negotiating and exchanging is the experimental task. There is no need to test whether participants believed the experimental manipulation for there is no misdirection. It would be accurate to say that, for the duration of the experiment, the participants actually lived and acted in the replica social structure. The structure just described is an exclusionary structure where A is high in power and the Bs are low. To control personal idiosyncrasies and balance payoffs, participants were periodically rotated across positions until each subject had negotiated from each position for a given number of times. Data were collected by position, not by participant. An alternative null exchange structure is identical to the one just described but for allowing the A position to exchange with all three B positions. In the null structure ET predicts equipower exchanges.

In this and other Elementary Theory experiments the whole of the theoretical model, except for its decision procedures, is realized in the replica. In the replica, the thinking of the participants takes the place of the model's decision procedures (i. e. resistance, etc.). As a result, ET experimental designs are both very simple to build and conduct, and highly accurate representations of the model being tested.

By changing the connectivity at the A position, power can be reversed. For example, let A exchange with three Bs and be required to exchange with all to profit from any of the exchanges. This is inclusive connection, and when inclusively connected, the Bs will exercise power over the A. Alternatively requiring the A to exchange in a given order with the Bs gives an ordered connection at A and, like inclusion, a decentralized power structure. Furthermore, types of connection can be mixed. Inclusion and ordering together produce greater power decentralization than either taken alone. Furthermore, inclusion and/or ordering can be combined with exclusion. As predicted above exclusion overwhelms the effects of inclusion and ordering.

Here is another example design. In the Fig. 8.2 social structure, three A positions can each reach and exchange with any of the four B positions. One B will always be excluded. The exchange relations may have sanction values like those of the power structure just discussed or may use 24-point resource pools as in power-dependence experiments. In either case, the As who are never excluded are predicted to be high power and the Bs, one of which is always excluded, are predicted to be low power. But for the number and connectivity of the positions, the Fig. 8.2 structure is theoretically

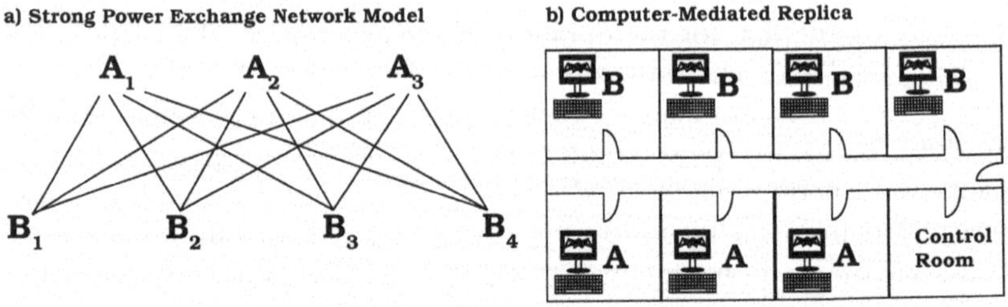

Fig. 8.2 A 7-position strong power networks and its computer-mediated experiment replica

similar to the exclusionary Fig. 8.1a structure. Exchange ratios soon approach the maximum favoring the As.

A coercive structure provides our final example. Shown in Fig. 8.3 is a centralized coercive structure with one C, the coercer, connected by coercive relations to five Ds, the coercees. The coercive relations are generated in the following way. C initially holds two negative sanctions each worth one to C and costing ten to any D who receives one. D initially holds ten resources each worth one if kept by D and one if received by the C. The structure has five of these coercive relations connected at C. C and D negotiate for the number of resources D will send to C in order to forestall C from sending a negative sanction to the D. When successful—from C's point of view—the threat alone will extract positives from any of the Ds. Since it is C's threat that extracts sanctions from the Ds, C's negative sanction is never sent. Therefore, C does not need five negative sanctions to extract value from the Ds.

As in exchange, the whole of the coercive theoretical model, but for its decision procedures, is realized in the replica coercive structure. In the replica, the thinking of the participants takes the place of the model's decision procedures (i. e. resistance, etc.). As a result, the coercive structure is both very simple to build and conduct, with highly accurate representations of the model being tested. The coercive structure just built is a *null* structure in which all the power being exercised is the result of threat grounded in the coercive relation. Applying resistance in the model predicts that each D will send approximately half of their resources to C.

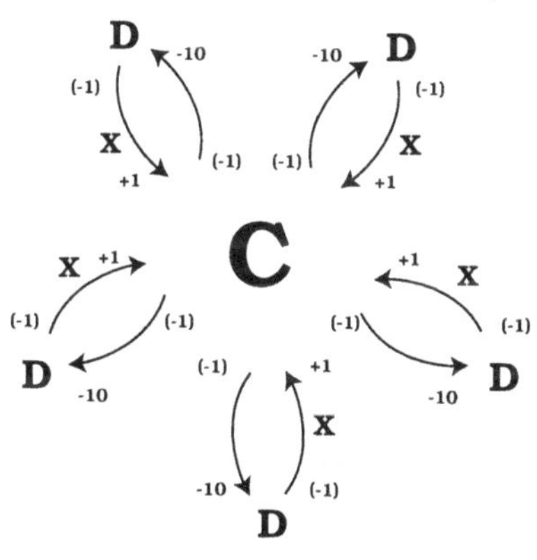

Fig. 8.3 A centralized coercive structure

To build a *strong* coercive structure, C must be motivated to send one or more negative sanctions to some of the Ds. That transmission produces a competition among the Ds to avoid receiving the negative. That competition is exactly like the competition among low power actors seen in strong exchange structures. Furthermore, the results of the two are similar. In this case rates of coercive exploitation are predicted to move to the extreme favoring the coercer.

Elementary Theory-driven experiments are unique among the theory-driven experiments discussed here, and, to our knowledge, unique among social theory-driven experiments, in its use of theoretic models as blueprints to build empirical replicas. When that method of experimental design is employed, it offers a number of advantages over other designs. For example, it is fundamental to theory-driven experiments that the experiment, as it is realized in the laboratory, exactly corresponds to the theory being tested. In Elementary Theory-driven experiments, that correspondence seems to us to be particularly easy to determine. It is straightforward to observe whether a model and its replica correspond or not and, if not, what to modify to establish that correspondence. Furthermore, when the modeling procedure has the flexibility to build an array of example social structures, corresponding replicas built using those models give uniquely diverse phenomena for investigation.

Looking to the future, there is an uncountable number of social structures worthy of study. Alternatively, new research could combine Elementary Theory models with formulations from other theories. Referring to theories discussed in this chapter, it would be instructive to people Elementary Theory model structures with some or all prosocial actors. Equally instructive would be to investigate collective action as a mixed power and influence problematic. It is known that power and influence are intimately related in many if not all social structures. Introducing influence into models for power structures would tell us much about the two phenomena and how they are related.

Conclusion

No one can say, with assurance, what Robert Merton's appraisal of the sociological research discussed in this chapter would be. It is hoped that he would be pleased that 'middle range theory' is alive and strongly developing. He feared sociology would balkanize and it has. By a recent count, this year there are 54 sections that will be represented at the American

Sociological Association meeting. Who can say what research, if any, reported in those sections is progressive? There is, however, no doubt that the sociological research reviewed here is progressive and reflects strong theory growth. By theory growth we mean the extension of the scope of theory through a series of experimental tests. For example, the research by R. Willer, Ridgeway and Nakagawa, and Melamed and Savage expand the scope of Status Characteristics Theory. That kind of theory growth is undoubtedly the single most important development in sociology since the classical era of Marx and Weber.

Given the significance of its growth, why has sociological knowledge not cumulated? The answer must be that theory growth and theory cumulation are not the same thing. Theory growth occurs when a theory forms a core from which the expansion of scope builds outward like the spokes on a wheel. To find how the growth parts are related, one must trace back through the hub that is the core theory. In theory cumulation, however, the growth forms an integrated whole. For example, in Elementary Theory, it is straightforward to infer directly from one type of structural connection to another—even one kind of social relation directly to another. Status Characteristics Theory has grown. Elementary Theory has cumulated. Not all growth needs to be cumulation. A field in science is progressive if either or both occur.

Looking across the sciences, it is evident that they are progressive because there is a special relation between theory and experiment. In progressive science the two work together. Some sciences that appear to be exceptions, like astronomy, are not exceptions. Astronomy, while not itself experimental, has been progressive since Newton integrated earth-bound trajectories and planetary orbits—thus integrating physics and astronomy. There are, however, exceptions. Plate tectonics and Darwinian biology are examples.

Is the special relation between theory and experiment is possible in sociology? Does experimental research suffer from limited 'external validity'. Nothing could be further from the truth. What, we may ask, is the external validity of survey research? The authors of this book have yet to encounter a theory-driven survey. Since there is no inductive method, and survey research does not test theory, it is restricted to inferences from sample to population. There are perfectly good reasons to gather that kind of information. But no one should confuse the transitory information produced by the survey with general knowledge. The survey infers from a smaller number of specifics to a larger number of specifics. When "external" means general knowledge, survey research has an external validity problem. Theory-driven experiments do not.

References

Bentley, A. F. (1949). *The process of government: A study of social pressures.* Principia Press. (Original work published 1908).

Berger, J., Cohen, B. P., & Zelditch, M., Jr. (1972). Status characteristics and social interaction. *American Sociological Review, 37*, 241–255.

Berger, J., Fisek, M. H., Norman, R. Z., & Zelditch, M., Jr. (1977). *Status characteristics and social interaction: An expectation states approach.* Elsevier.

Fisher, R. A. (1935). *The design of experiments.* Oliver and Boyd.

Gervais, W., & Norenzayan, A. (2012). Analytic thinking promotes religious disbelief. *Science, 336*(6080), 493–496.

Hage, P., & Harary, F. (1996). *Island networks.* Cambridge University Press.

Lucas, J. W. (2003). Theory-testing, generalization, and the problem of external validity. *Sociological Theory, 21*(3), 236–253.

Melamed, D., & Savage, S. V. (2016). Status, faction sizes, and social influence: Testing the theoretical mechanism. *American Journal of Sociology, 122*(1), 201–232. https://doi.org/10.1086/686943

Oliver, P. (1980). Rewards and punishments as selective incentives for collective action: Theoretical investigations. *American Journal of Sociology, 85*, 1356–1375.

Olson, M., Jr. (1965). *The logic of collective action: Public goods and the theory of groups.* Harvard University Press.

Popper, K. (1959). *The logic of scientific discovery.* Harper & Row. (Original work published 1934).

Ridgeway, C. L., & Nakagawa, S. (2017). Is deference the price of being seen as reasonable? How status hierarchies incentivize acceptance of low status. *Social Psychology Quarterly, 80*(2), 132–152. https://doi.org/10.1177/0190272517695213

Simpson, B. (2006). Social identity and cooperation in social dilemmas. *Rationality and Society, 18*, 444–470.

Weber, M. (1968). *Economy and society.* University of California Press. (Original work published 1918).

Willer, R. (2009a). Groups reward individual sacrifice: The status solution to the collective action problem. *American Sociological Review, 74*(1), 23–43.

Willer, R. (2009b). A status theory of collective action. In S. R. Thye & E. J. Lawler (Eds.), *Advances in group processes* (Vol. 26, pp. 133–163).

CHAPTER 9

Theory Toolbox

As seen in Chap. 6, theory is the method that sciences use to explain and predict (Popper, 1934; Willer & Emanuelson, 2021). The focus of this chapter is historical explanation, employing a toolbox of experimentally tested theories. In building social theory, the first step is to construct a diagram. Diagrams are the social structures in which activity takes place. Taken alone, diagrams are static. Diagrams become dynamic models when set in motion by their mechanisms. Decision procedures are the mechanisms of the model. The second step is to people the positions of the diagram with the decision procedures by which actors act. Thus, a model is a diagram with positions and actors that are filled with mechanisms. Mechanisms employed in social theory include:

1. the rational choices agents make in game theory,
2. the decision procedures of actors in Elementary Theory, and
3. the parametrically rational linkage from status to influence in Status Characteristics Theory.

As seen in Chap. 8, once a model is built, the next step is to build its replica which is the design of the experiment. The replica is the wholly empirical representation that corresponds to the diagram of the abstract model. The decisions and actions of the humans acting in the experimental replica take the place of the mechanisms of the model. The model is

D. Willer, P. Emanuelson, *The Logic of Theoretical Sociology*,
https://doi.org/10.1007/978-3-031-91709-7_9

analytic. The replica is synthetic. The dynamics of the model are predictions for the replica. Having run the replica, when its dynamics correspond to those of the model, the test has succeeded. If this is a test of a newly developed replica, and the dynamics of model and replica correspond, the scope explained by the theory has been enlarged.

Historical and contemporary applications of theory reverse those steps. The aim in explanation is not to produce the phenomenon in the laboratory, but to fit theoretical models to the facts given in the historical record. Our point of departure is a number of experimentally tested social structural models each with known properties and dynamics. That these models have been experimentally tested lends confidence in their explanatory power when historically applied. The second point of departure is to select the historical events to be explained. For example, Marc Bloch in his *Feudal Society* (1961) told of successive invasions and besiegements of European society by Huns, Islamics and Scandinavians. He saw these invasions as conditions favoring the development of feudal social structures. Bloch's inference, that he shares with Simmel (1955, p. 97), is supported, in part, by experimental investigations linking intergroup conflict with intragroup cohesion (Bornstein, 2003; Benard, 2012). Beyond intragroup cohesion, explaining why the cohesion that developed was produced by feudal social structure and not by some other structure demands that further theory be deployed.

The purpose of applying theory is to explain historical and contemporary events. Historical and contemporary events are complex, as are all events outside controlled laboratory conditions. How are complexities to be explained given that theory, tested in simple lab conditions, is itself simple? The problem is not unlike that of J. S. Mill in that both are faced with a complex world of blooming, buzzing confusion. Unlike Mill complexity is not explained with regularities discovered by cutting up the world into component parts. Mill's methods do not find regularities nor can any other supposed inductive method. In fact, there is no inductive method and regularities may or may not be "out there" to be found. Instead, order is imposed by theory. How does simple theory impose order on empirical complexity? To impose order on historical events calls for not one theory but a set of theories. Following Hedström and Udehn (2009, p. 31), let's call that set a theory toolbox.

To move from the simplicity of lab tested theory to the complexity of historical events models are pulled from the toolbox and compared to structures and dynamics of historical events. As experimental investigations have accumulated, a toolbox becomes full of multiple modeled structures with known properties. Comparing these models to historical events frequently uncovers historical properties and dynamics not previously thought to exist. The purpose of employing a toolbox of theories is most emphatically not to offer a shortcut to understanding historical and contemporary events. Using a toolbox for explanation demands more not less effort. That effort includes both gaining an understanding of an array of experimentally tested theories and acquiring knowledge of the historical events to be explained. Using a toolbox to form an historical explanation is a synthesis, not a shortcut.

Theory-driven tests of social theory over the last decades of the twentieth century and the first decades of the twenty-first have produced an array of experimentally tested social structural models. Many of which will be employed here. In spite of these riches, it would be unreasonable to expect that our theory toolbox would necessarily contain all the models needed for any given historical explanation. Fortunately, the theory toolbox is not static. New models can be added. Furthermore, when a model needed for explanation is missing, models already applied may give clues to building that new formulation.

Here is an example of what was done when theory was missing. The first historical explanation offered here is the rise of the political state. The state developed through a complex step-by-step process that began with an influence-based chiefdom and proceeded through coercion-based chiefdom to a coercion-based state. Having arrayed models for social structures in temporal order, it was evident that there was theory missing that would connect influence-based to coercion-based chiefdoms. For modeling of this or any historical process to be complete, it is not enough that the models be arranged in a plausible order. They must fit together with matching configurations like the pieces of a puzzle. More precisely, the end conditions of each modeled structure must fit into the initial conditions of the next. But, when building the sequence of models, it became clear that further theory was needed to connect two steps of the process. The end conditions of the influence-based chiefdom did not fit the initial conditions of the coercion-based chiefdom. Since the fit was not there, the theories currently in the toolbox could not explain how the influence-based chiefdom evolved into the coercion-based chiefdom.

Thinking again, the authors remembered Marx's comment that societies change when a new structure evolves within the old society. As he put it

> No social order is ever destroyed before all the productive forces for which it is sufficient have been developed, and the new superior relations of production never replace older ones before the material conditions for their existence have matured in the framework the old society. (1859/2009, pp. 20–21)

The modeling process led to a recognition that status-based chiefdoms were warlike and that wars were dividing the chiefdom into two strata: the aristocrats who fought and the commoners who worked. A model in which protection emerged—in which a higher status stratum offered protection to a lower status stratum. Like protection offered by the Mafia, the protection came for a price. The model for 'protection for a price' is a structure of indirect coercion. (See below.) This new model for indirect coercion was formed and successfully tested experimentally. Now final conditions of the model for influence-based chiefdom matched the initial conditions of the indirect coercion model. And the final conditions of the indirect coercion model fit the initial conditions of the coercive chiefdom model. The new modeled process completed the theory for the development of the state. Importantly, the theory toolbox is not and should not be static.

As sociological theory grows so does the toolbox, increasing the ability to explain historical and contemporary events. Let's now turn to three example historical processes and structures: (1) how the political state developed, (2) how European feudalism devolved, and (3) how the equilibrium between the Mongolian tribal configuration and Chinese state was balanced. The models applied for each are intended to explain, at least in part, the dynamics of the respective social structures.

Development of the State

As modeled here, the state develops through four steps, the first where influence is exercised by those of high status over those of low status and three further steps where various forms of coercive power are centrally exercised. Elementary Theory (ET), previously presented in Chap. 7, is used to model the three power structures. For those models the authors will employ ET's formulations for actors and structures. As will be remembered, beyond structural position, actors' actions are governed by values and beliefs, and either or both of these must be specified for each actor. Borrowing from Marx, however, it is efficient to treat both as reflections of the actor's conditions of existence. As quoted earlier, "It is not the consciousness of men that determines their social existence, but their social existence that determines their consciousness" (Marx, 1859/2009, p. 20). It follows that the simplest model is one in which every actor's values are given by the actor's position in the structure while others' values and the other reflective actor's beliefs are the structure itself.

The first of the models—the model for structurally induced influence—is an exception for it does not employ ET. As currently developed, ET has no formulations for influence. Therefore, formulations for structurally produced status and for status induced influence are pulled from the theory toolbox.

Status-Lineage

A status-lineage is a descent group in which statuses are inherited. Whereas status-lineages, like building blocks, can be embedded in a larger power structure, a status-lineage *chiefdom* is an autonomous status-lineage structure. The individuals of any status-lineage can trace or remember their own and other's genealogy. Beyond memories of individuals, there may be specialists in genealogy who, sitting in panels, judge an individual's descent (Malo 1898/1903, p. 254). While inheritance of status may be on the male side, female side or jointly on both, all three are patriarchal (Keesing, 1975, p. 63; Aberle, 1961; Schneider, 1961; Parkin, 2004). Though the theories employed here can cover all three descent types, the focus will be on the patrilineal form because it is simplest to diagram and interpret.

Status Structure of the Chiefdom

Figure 9.1 diagrams a simple patrilineal status-lineage structure in which lower numbers designate higher statuses. "F" is the founder, number 1 is his first-born son, number 2 the first-born son of that son, and similarly tracing the senior line to the right. Tracing back to the left and down, number 8 is the founder's second-born son, number 9 is number 8's first-born son and 10 his second—and similarly for the whole of the diagram. Larger diagrams follow the same logic. For simplicity each father is shown with two sons with the first son inheriting the father's position. The result is a very large status difference between first and second son.

To find these statuses in the lineage structure Let's dip into the theory toolbox and pull out Hage and Harary's (1996) method that assigns a

Fig. 9.1 Simple, Patrilineal Status-Lineage Structure

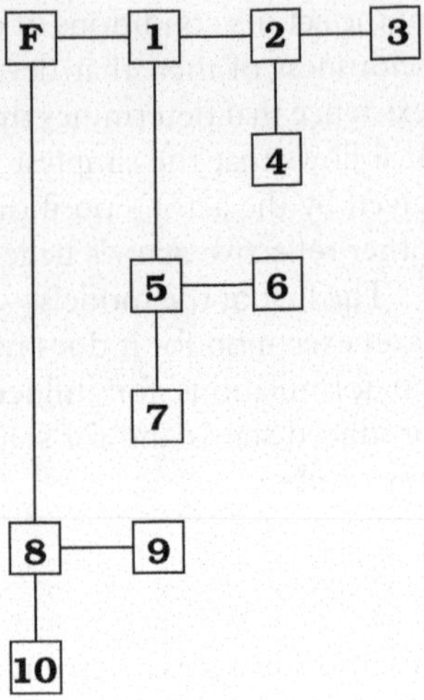

unique status to each position in the lineage. As traced above, the Hage-Harary method begins at the founder and counts across the senior line to the right, assigning statuses in order. Having reached the end of the senior line, it traces back to the left assigning statuses down each branching as they occur. Beyond assigning statuses, the Hage-Harary method uncovers two embryonic cadet lines of 5 → 6 and 8 → 9 → 10.

Some qualifications are in order. Status-lineage structures are broadly found on every continent inhabited by humans including the Han Chinese, the populations of ancient Greece and Rome, and some, if not all, tribes invading Europe as Rome fell. Whether status is assigned in a way consistent with Hage-Harary in all these instances cannot be claimed. Nevertheless, the Hage-Harary method for assigning status is logically consistent, and, to the authors' knowledge, is the only method extant for assigning status in status-lineage structures. But it should be mentioned that no proof is offered by its authors that any status-lineage structure assigns status consistent with their method. The generality of the Hage-Harary method, however, is not crucial to our explanation. The theory being formulated here can be applied using any method that consistently assigns status in the lineage structure.

Status and Influence

Returning to the theory toolbox, let's now deploy the 'Status and Factions' theory of influence (Melamed & Savage, 2013a, b, 2016). Status and Factions theory (SF) contains Characteristics Theory (SCT) as one link between status and influence and formulations for factions as a second link. According to SF theory, the extent to which alter is influenced is the result of both the status and number of egos. Unlike SCT, SF is not limited to the dyad. SF can be applied to infer influence in status structures like that of Fig. 9.1.

Since influence follows status, the position labeled 1 in Fig. 9.1 can influence position 2 and any other higher numbered position. Since influence also follows number, positions 2 and 3 can influence any lower status position or pair of positions—and similarly through the structure. It follows that because Fig. 9.1 is a tree with no status reversals, it is a well-ordered status structure. Inferring from Simpson et al. (2012), because it is well-ordered, the direction of influence is unambiguous, and the structure can solve collective action problems.

The status-influence chiefdom, the simplest form of chiefdom, is a system of privilege whereby the centralization of influence assures that style-of-life follows status. For example, the chief's largely empty claim to being generous allows chiefs to profit from their systems of redistribution (Sahlins, 1958; R. Willer et al., 2012). To be explained later is why status-influence chiefdoms are warlike. Here let's, following Kelly (2000), that unlike the more egalitarian clan, the status-lineage chiefdom is well enough organized that it can wage war. As Kelly (2000) found, war and stratification go hand in hand. He found that war was almost wholly absent in egalitarian, segmented societies (p. 146), whereas in ranked structures such as chiefdoms, war, not peace, is the normal condition (see also Carneiro, 1970, 1991, 1998, 2012; Earle, 1991b; Marcus, 1998). With the addition of constant war, the end condition of the status-based chiefdom matches the initial conditions for the next step, the indirect coercive chiefdom.

Indirect Coercive Chiefdom

Unlike modern war that is highly destructive to winners and losers alike, prehistoric and ancient war could be very profitable indeed, if only for winners. For example, Rome increasingly filled its laboring class by capturing and enslaving people through war (Weber, 1896/1988). The contrast is due, at least in part, to the very destructive power of modern war with its diverse ways of exploding things and the very slow rates of production in ancient societies. Thus, it should not be surprising that war was the normal condition among Greek city-states (Hansen, 2006) as well as among status-linage chiefdoms.

The indirect coercive chiefdom is a social structure in which a high-status grouping, the aristocracy, offers 'protection' to a lower status grouping, the commoners (Emanuelson & Willer, 2018). Protection means that there are negative sanctions external to the aristocrat—commoner relation that the aristocrats can block so that they do not negatively affect the commoners. The result is the power relation of indirect coercion. War between and among chiefdoms produces opportunities to provide protection and exercise power within each. Indirect coercion occurs in the chiefdom when the aristocracy demands support from commoners in order to defend the commoners against hostile outsiders real or imagined. In anthropology the terms 'aristocrat' and 'commoner' designate status groups with differing lifestyles. In the model now being developed, the two status groups are wholly defined by their place in the indirect cohesive structure. The diagram shown in Fig. 9.2 is a simple case, having one aristocrat and multiple commoners.

Unlike direct coercion where coercers threaten coercees, the aristocracy benefits from indirect coercion, but not by threatening to sanction the commoners. Instead, they threaten to withdraw protection. As a result, the exercise of indirect coercion does not require the concentration of the

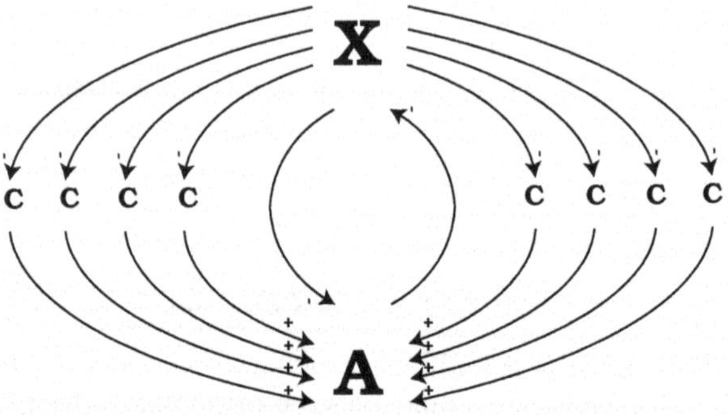

Fig. 9.2 Indirect Coercion of Commoners by an Aristocrat

means of violence found in the coercive chiefdom and modern states. It is sufficient that relations between a chiefdom and its neighbors be hostile so that the aristocracy can offer to defend the commoners—for a price. Experiments by Emanuelson and Willer (2015) compared rates of exploitation between structures of indirect and of direct coercion and found them to be of the same order. The two forms were both capable of the exploitation form of power. It seems doubtful, however, that indirect coercion, even when paired with influence, could be as effective as direct coercion at the control form of power.

Moving from influence to power, SF is returned to the toolbox and Elementary Theory (ET) (Willer, 1984; Willer et al., 2014), with its flexible modeling procedure, is used to model the relations and structures of three types of the chiefdom and the state. Figure 9.2 is a diagram for a small indirect coercive structure. The diagram becomes a theoretic model through the addition of a mechanism. The mechanism is generated by peopling the positions of the structure with rational actors which act from the positions of the diagram. This is a 'reflective' model where actors acquire their values and beliefs from their positions. Here and below, ET models used in this explanation have known properties that have been tested in the laboratory.

With the development of indirect coercion, the chiefdom structure undergoes a fundamental transformation. Now an aristocracy has fully emerged with a new status ranking, that of warrior caste. That ranking stands between the chief who is first among equals and a lower status group, the commoners (Flannery & Marcus, 2003). Becoming a member of the warrior caste is to be upwardly mobile in a society in which statuses had previously been fixed at birth. With the development of the warrior caste, the end conditions of the indirect coercive chiefdom fit the initial conditions for the third step of chiefdom, the coercive chiefdom.

From Coercive Chiefdom to Complex Chiefdom

Through direct coercion, property in land came to be privatized (Hudson & Levine, 1996, 1998; Emanuelson & Willer, 2018). For example, in Hawaii, under Chief Kamehameha, commoners were coercively separated from their land (Kirch, 2010, p. 184ff). In England, William claimed title to all land after the successful invasion of 1066. While Kamehameha and William sought to own all land, in the absence of an administrative structure that could extract taxes from so many commoners, some land came to be owned by members of the warrior aristocratic caste. In Hawaii, the chief and aristocracy denied that commoners were related to them in any way—they were excluded from the status-lineage structure. The same effect was produced by the Church in England where the greed for land pruned the extended lineage down to two generations (Goody, 1983, p. 1983). More generally, to extract value from land, as discussed in the section on feudal social structure below, chiefs and kings came to use land as payments for services in war. As a result, the aristocracy was transformed into landed gentry. (See also Earle, 1991a.)

According to Kirch (2010), the then chief of Hawaii's Big Island collected tribute by traveling from place to place collecting tribute at each stop—as did the kings of European feudalism. Kings and chiefs were impelled to move because they and their entourage would soon use up locally available food. Had the king or the chief an administrative staff, they could have exercised power-at-a-distance to rule and feed the staff instead of traveling from place to place to exercise power directly (Willer, 2003).

Coercive chiefdoms grow into complex chiefdoms in the following way. While the status-lineage is itself size-limited, status-lineages can, like building blocks, contribute to the size of a resulting complex chiefdom (Barfield, 1990, p. 153ff). Imagine a series of Fig. 9.3 coercive structures connected at the chief positions. As put by Flannery (1999, p. 4), "Chiefdoms grow complex by taking over their neighbors, demoting the latters' chiefs to

Fig. 9.3 Direct Coercion of Commoners by an Aristocrat

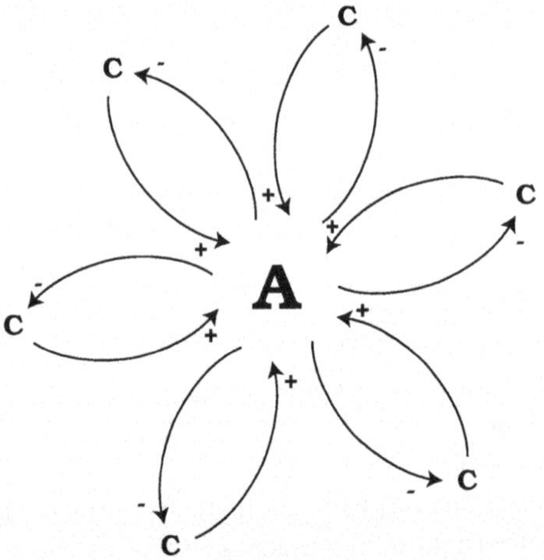

subchiefs or replacing them." In other words, a complex chiefdom consists of a number of subordinated chiefdoms led by a paramount chief and warrior caste. The complex chiefdom has the advantage of numbers in warfare relative to unallied adjacent chiefdoms. As a result, complex chiefdoms tend to grow by incorporating the chiefdoms around them.

In complex chiefdoms, as in simpler coercive chiefdoms, the exercise of coercive power is limited to power over adjacent positions. Since coercive power does not extend through the structure, parts are held together by familial relations among those at the chiefdom level. For example, in the Trobriand Islands, the paramount chief takes many wives each of whom "is always the sister or some relation to the headman of the subject village" (Carneiro, 1981, p. 77). In parts of Asia and Africa, multiple marriage by the head of state was a common practice even when monogamy was the rule for the larger society. It follows that the cohesion between the parts of the complex chiefdom is weakest at the chiefdom level and is weaker than the cohesion within any of its parts. This point of weakness has been noted by many anthropologists. Marcus and Feinman (1998, p. 93) point out that the status-lineage parts of primary states are also more stable than relations at the top of the large polities of which they are a part.

For stability, structures as large or larger than complex chiefdoms must solve the problem of power-at-a-distance. Prior to that solution, as noted by Anderson (1994), Wright (2006) and others, cycling between coercive chiefdom and complex chiefdom structures should be expected. While cycling signals that state formation is imminent, cycling will continue until the problem of extending power beyond adjacencies is solved.

Status-Rivalries and War

The process of state development that begins with status-lineage and ends with the state is driven by ongoing conflict. Why are chiefdoms and early states warlike? To answer let us return to the toolbox to pull out the results of experimental studies that show the relation between downward mobility and solidarity. They show that chiefdoms are warlike because there is an inherent instability in their structure that leads them to war. That instability is the downward mobility that is an integral part of the status-lineage structure. The authors will first show that status-lineages are beset by downward mobility and then relate how, as a result, that mobility leads to war.

That downward mobility is inherent in chiefdoms can be seen by tracing the shaded position across the three time periods of Fig. 9.4a–c. In Fig. 9.4a the shaded status position, the second son of the founder, has the second ranked status position. At a later time, as shown in Fig. 9.4b, after his older brother has had two sons, the status rank of the second son of the founder has declined to fourth. As time goes forward, the older brother's sons have had sons of their own and, as a result, the second son of the founder's status rank has declined to eighth. Except for the first-born sons who are on the senior line, *all* in the status-lineage are plagued by downward mobility. That downward mobility is quite general. Note the ranking of any position across any of the time periods. For example, in Fig. 9.4b, the second son of the founder has had a son of his own. That son's status is initially fifth. Later, in Fig. 9.4c, his status rank has declined to ninth. Goldman (1960) attributed status-rivalries to inequality: "The psychology of status is such that inequality provokes rivalry, if not through the entire

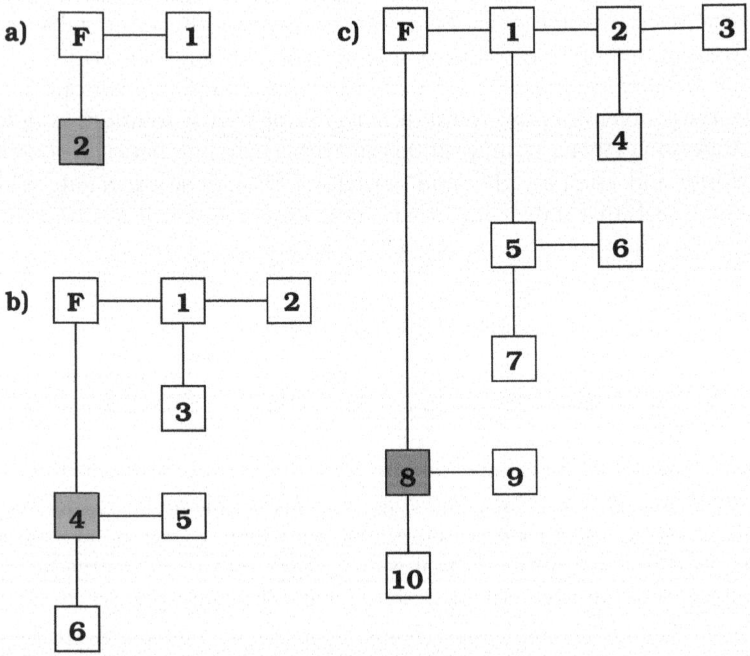

Fig. 9.4 A Status Lineage with Downward Mobility

society, then surely in its upper status ranges" (p. 691). The prime source of rivalries is not low status which is static but loss of status which is dynamic. We do not deny that inequality is a source of rivalry. What we suggest is that decline of status is the stronger effect. Note that, when commoners are ejected from the status structure, status rivalries will be, as Goldman noted, limited to the higher statuses.

Status-rivalries disrupt lines of influence and thus the ability of influence-based chiefdoms to organize activities. Experiments by Barclay and Benard (2013) show that disrupting a status ordering disrupts its lines of influence and thus its ability to organize. Experiments by Kilduff et al. (2016) show that status disagreement leads to withdrawal of contributions and lower group performance. Pettit et al. (2010) show that the effect of status loss was greater than the effect of an equivalent status gain (see also Benard, 2012; Benard & Doan, 2011). That status loss has greater effect than status gain agrees with Kahneman and Tversky's (1979) Prospect Theory. The effect of downward mobility is by no means wholly subjective. Objectively, loss of status goes hand in hand with loss of wealth benefits that flow to those of high status.

Given status-rivalries, it is evident that the path of development across chiefdoms is not a superhighway. Instead, it is better thought of as a narrow path bordered on both sides by precipices down which any chiefdom could fall. And many did fall, due to status-rivalries.

Why does downward mobility make chiefdoms warlike? The answer is that, through war, chiefs and their aristocracies can counter the disruptive effects of status-rivalries and maintain their privileged positions and their ability to organize activities. An array of notable theory-driven experimental studies relates external conflict to internal interest situations. Experiments by Bornstein (2003) and Bornstein et al. (1990, 2002) demonstrate that the threats of intergroup conflict interact with group payoffs to reduce free riding and increase group cohesion. Puurtinen and Mappes (2009) found that intergroup competition resolves collective action problems by increasing group cooperation. Further studies show that external threats increase support for leaders (Gavrilets & Fortunto, 2014; Van Vugt et al., 2008). Pursuing war to increase solidarity and support of leaders is a policy that is still practiced. As documented by R. Willer (2004), U.S. President George W. Bush, when facing rapidly declining support, demanded higher taxes and increased powers to pursue war justified by imaginary threats of mass destruction from Iraq. When commoners are denied family status in the status-lineage, they are not downwardly mobile, they are already at the bottom. But having little or no influence on polity policy, it matters not whether they experience status rivalries. Since policy is formulated by chiefs and members of the warrior caste, downward mobility and associated status-rivalries still have their effect.

Social Structure of the State

For modeling the structure of the state, let's keep at hand the model for direct coercion and pull out from the toolbox ET's model for hierarchies (Willer, 1987) and ET's model for power-at-a-distance (Willer, 2003). The first model explains that actors in hierarchies who are not fixed in their positions are subject to power exercised by superiors. The second model traces power beyond adjacencies and shows that that exercise requires structural power conditions between each pair of levels. Power-at-a-distance and hierarchical power are the two organizational conditions that distinguish the state from simpler coercive structures. "State" means a polity with centralized coercion administered by a purposefully designed organization. Contrary to Weber (1918/1968), not all modern polities should be designated as states. Only the most advanced states, and only over the last century and a half, have eliminated private vengeance. Only they can be said to have concentrated the means of violence.

Wright (1977), Redmond and Spencer (2012), and Flannery and Marcus (2012) agree with toolbox models that the form of the state is distinguished from the complex coercive chiefdom by the division of administration into distinct tasks. This division may be quite simple. For example, in the Roman Empire, prior to its late bureaucratization, only two functions were separated: civil and military. The emperor maintained power centralization by playing the two off against each other. When administration is divided into separate tasks, each task is carried out by a bureau, that is to say a governmental office. Having multiple bureaus, the state is bureaucratic. The establishment of multiple bureaus is also central to Weber's (1918/1968) formulations for bureaucratic organization (See below). Elias (1994) investigated the development of the centralized French state that evolved out of decentralized feudalism. He suggests that developing division of administrative tasks of the evolving state reflects an underlying developing division of labor in the society of that state (259ff).

Flannery and Marcus' (2012) ethnohistorical investigation of early states suggests that states do not arise one at a time. They evolve in groups of two or more.

> In the four cases we examined, not one kingdom was the offspring of a rank society that simply got bigger. There is apparently no social steroid that can trigger that kind of growth. Instead, all four kingdoms arose through the forced unification of a group of competing rank societies. (p. 362)

Flannery and Marcus' joint state evolution is a worthy hypothesis, but one we cannot test here.

Since states arrived before money, officeholders of the earliest states were necessarily paid in kind by some combination of plunder and benefice. A benefice is payment attached to the office, of land, usually including unfree or semi-free labor tied to the land (Weber, 1918/1968). Following Weber this may be called "patrimonial rule" where officials, recruited from the aristocratic warrior caste, are endowed with incomes in land. It is a relatively simple structure that has "offices vested with powers and authorities" that Kurtz (2012, p. 69) insists is central to power exercise in the state.

The founding of patrimonial rule is a straightforward extension of the ruler's household. Unlike modern bureaucracies where the ruler's power over the organization is based on a system of promotion in the hierarchy, in patrimonial rule power is based on the ability of the ruler to hire and fire officials—and upon the inability of officials to reciprocate. Unlike modern bureaucracy where positions are filled from below, patrimonial rule is characterized by "peer recruitment". "Peer recruitment" means that office holders are moved horizontally from status positions outside the organization to corresponding status positions inside. Given peer recruitment there is no *system* of promotion. As long as office holders do not own their positions, however, they can be moved between positions at the whim of the ruler. Nevertheless, as soon as patrimonial rule is established, a struggle for ownership of the benefices begins (Weber, 1918/1968, p. 1042ff) and sometimes succeeds, as with the Magna Carta.

Theory Toolbox I

The toolbox employed in this section of the chapter contains experimentally grounded theories and models drawn from them. The models take a "structurally individualistic" form (Udehn, 2001) in that social structures are explicitly drawn and are made dynamic by actions of actors whose values and beliefs are acquired from the structures. The models drawn from the toolbox form a unified whole because the model's social structural conditions that were end conditions of each step corresponded to the initial conditions of the model of the next step. Explaining when a model is pulled from the toolbox and when a used model is returned, highlights how multiple theoretics were applied in concert. Importantly, no micro-macro problems in applying experimentally tested theory to large social structures have been encountered.

None of the theories and models deployed here sees ruling groups as planning transitions between social structures. None assumes that formation of a state was the goal of any historical actor. The histories to which the theory toolbox applies were certainly made by the people who lived them, but not under conditions chosen by them. Turning to the rise of Feudalism, let's bring along the theory toolbox, just used for application to the next problems of explanation.

The Transition to Feudal Social Structures

Marc Bloch (1961a, b) opened his brilliant *Feudal Society* by recounting the external pressures on western Europe that accompanied the fall of Rome in the west. The invasions and besiegements he cited were the Hungarians from the east, the Islamics from the south, and the Scandinavians from the north. As already seen in this chapter, there is experimentally tested social theory that supports the inference that external pressure leads to internal order. Feudalism, in defending against external threats, did impose internal order. But why was feudalism fostered by external pressure and not some other social structure? Answering that question will require a closer look at the transition from the structure of the late Roman Empire to early feudal social structure.

European feudalism was grounded on a social relation quite unlike exchange, the relation on which capitalism is based. Exchange relations, while not unknown, were unusual to feudalism and, at times, wholly absent (Finer, 1997). Why since the fall of feudalism has the realm of exchange expanded to fill everyday life? North and Thomas (1973) claim that exchange relations replaced other social forms because they are more efficient. But efficiency alone cannot explain why exchanges became all-pervasive. If their view is correct, we also need to know what institutional mechanism selected for efficiency. On that issue North and Thomas are silent.

Contrary to North and Thomas, the mechanism that selected exchange is unrelated to efficiency. Today exchange dominates social life, not because it is more efficient, but because the modern state has monopolized the employment of negative sanctions that are necessary to coercion and conflict. What does state monopolization of coercion and conflict have to do with the dominance of exchange? Let's look back at Fig. 7.2 of Chap. 7 where the three basic types of social relation, exchange, conflict and coercion, are displayed. The modern state monopolizes negative sanctions of coercion and conflict. Since it does, by default, the social relation that remains is exchange. With the centralization of negative sanctions into the structure of the modern state, exchanges expanded to fill the social space.

The Fall of Rome: The Transition from Strong to Null Coercive Structure

Living in a capitalist society where economic growth is the normal condition, it can be difficult to grasp how and why the society of classical antiquity in western Europe collapsed backward into feudalism. Nevertheless, there has been no lack of ingenuity in finding causes for the collapse. Alexander Demandt is notable for having found 210 causes. Of that number, for the theory toolbox, one stands out: the transition from slavery to serfdom. For Marx, this transition marks a fundamental change in how the superstructure squeezed surpluses out to the base. Though different in many ways, slavery and serfdom are similar in that both are means of coercive exploitation. The two differ, however, in the functioning of their coercive structures, a difference that results in very different rates of coercive exploitation.

As explained in Chap. 7, all coercive relations are power relations. It follows that any social structure composed of coercive relations is a power structure. If that power is produced by the relations alone and not by the structure, it is a null coercive structure. Null coercive structures have medial rates of exploitation. If that power is produced by relations and structure together, it is a strong coercive structure. Strong coercive structures have high, even maximal rates of exploitation.

Here is how the two structures function. Since both are coercive, both work by force threat. Resistance equations have been used to determine what threat transfers what proportion of value from coercee to coercer. In historical coercive structures such as serfdom or slavery, the lord or master (respectively) faces practical limits to the size of the threat. A dead serf or slave produces nothing, and any injury reduces productivity. In null coercive structures, these practical limits of coercion result in medial rates of exploitation.

The strong coercive structure is an analogue of the strong exchange structure. As will be remembered from Chap. 7, in strong exchange structures power is exercised by one or more high power positions. High power positions are connected to a surplus of low power positions that are exchange alternatives. The low power positions have no surplus exchange partners. As a result, some number of low power positions will be periodically excluded from exchange. Marx called those excluded under capitalism the 'reserve industrial army'. Structural power is exercised because, seeking to avoid exclusion, those in low power positions make increasingly better offers to the high-power position(s).

In strong coercive structures, structural power is exercised by the coercer which routinely sends the threatened negative sanction to the coercee who offers the least to the coercer. In slavery, the coercees offering the least to the coercer are the least productive slaves. Structural power is exercised because coercees, facing negative sanctioning, make better and better offers to the coercer. In slavery, slaves seek to avoid being the least productive. Experiments conducted by Willer (1987) found that strong coercive structures had distinctly higher rates of coercive exploitation than did null coercive structures. Importantly, the rates were distinct even though the size of threat was exactly the same in the two structural types.

Experimentally tested theory implies that Roman slavery was a strong coercive structure with high rates of exploitation while early medieval serfdom was a null coercive structure with its lower rates of exploitation. Working from historical documents, Weber reached a similar conclusion. He explained that:

> the Roman plantations depended on a slave market which received regular supplies of human material. But what if the market failed…that is just what happened. (1896/1988, p. 398)

According to Weber, Roman slaves did not reproduce themselves. Rome's regular supply of slaves was the result of the geographic expansion of Rome. In fact, the supply was particularly bountiful after Caesar's successes in Gaul. But under Agustus, expansion slowed and increasingly took on the form of slave raiding (Weber, 1896/1988).

Whatever policy was pursued by Rome, any state where the economy is dependent upon acquiring humans for slaves by expanding its borders is ultimately doomed by the linear-square law. According to that law, for any given time interval, the rate at which booty, including humans acquired to be slaves, is determined by the length of the state's border. By contrast, the demand for slaves is given by the area of the state. As the state's size increases slave supply, as determined by the length of the border, increases lineally while demand for slaves, as determined by the area of the state, increases to the square. Eventually, demand must outrun supply and, when it does, the system begins to collapse. In the case of Rome, the collapse is the transition from the slavery of strong coercive structures to the serfdom of null coercive structures.

In fact, the Roman Empire in the west was almost unique among ancient states in the extent of its dependence on slaves acquired by expansion. More common patterns of coercive exploitation that followed from expansion include (1) reducing new populations to serfdom and/or (2) taxing newly acquired territories, while leaving their social structures intact. For example, when Rome expanded into Greece, it left the social structures of the city-sates intact, but taxed them, with the local aristocracy being responsible for tax collecting (Hansen, 2006). Either pattern need not expand to be stable. But Rome in the west was stable only as long as it expanded *and* as long as that expansion was at an increasing rate. For some time, Rome's need to expand at an increasing rate was masked by Caesar's extraordinary successes in the west and by its expansion into the more densely populated east from the lower density west.

By contrast, feudalism was stable without expansion because its serfs reproduced themselves. The fact that serfs reproduced themselves under feudalism is known by the lack of extensive slave raiding. Furthermore, the much simpler superstructure of feudalism, in comparison to that of the Roman Empire, indicated that exploitation under serfdom was at a substantially lower rate than under slavery. The transition from slave to serf was crucially important for the transition from Roman Empire to feudal Europe. Nevertheless, other social structural changes worked hand-in-hand to hasten the collapse of the exchange economy and the Roman state. With those changes came the rise of feudalism.

Preconditions of Feudal Society

Long-term changes in the Roman social structure worked together with the fall of the supply of slaves to hasten the collapse of Rome's money economy. As the money economy collapsed feudalism was the residue.

These changes are called 'privatization'. Privatization consists of the growth of social structures within the state that are owned or controlled by a status group that is powerful enough to remove those structures from the state's tax base. As explained by Hudson, referring to ancient empires:

> Rulers regularly annulled their economy's debt burden so as to limit the privatizing of once-public resources. But ultimately, private wealth grew strong enough to undercut royal power…Economic power was turned into political power, abolishing royal oversight and removing the richest property from taxation. Oligarchies proceeded to expropriate poor cultivators and turned them into serflike dependents. (2018, p. 3)

The growth of large, privatized social structures was perhaps the single leading cause of the decline of ancient empires. Since the bronze age in the Mideast, the cycle of state decline after a long period of growth was often produced by privatization and the social structures used for the privatization were frequently very large religious organizations (Edzard, 1998; Buccellati, 1998; Dandmayev, 1998). These religious organizations were controlled by the oligarchy.

The effect of privatization on states is twofold. In addition to removing the means of production, that is to say land, from the state's tax base, it removed a portion of the population from military obligations. What did the ancient state need to reproduce itself? The needs were two: money from taxes and a supply of young men for the military. The Roman Empire in the east was not based on slavery. Instead, it worked to preserve the economic position of small landowners precisely because their young men were a needed military resource. Under that policy, unknown in the west, the Roman Empire in the east did not need to expand to be stable.

In the west the growth of *latifundia*, large, privatized estates owned by oligarchs contributed to Rome's decline (Hudson, 1995, 2023, p. 5). As quoted by Hudson, at least one Roman historian, Pliny recognized the threat of privatization. In Rome, the politically strongest status group, the senatorial oligarchy, had the political power to privatize and, in doing so, built the *latifundia*. As the *latifundia* grew, Rome's tax base shrunk. As elsewhere in antiquity, beyond reducing the tax base, privatization also eliminated the class of small landholders who were the source of recruits for the military. As the barbarian pressures increased, the army increasingly reproduced itself.

Turning to the internal structure of feudalism, it is helpful to remember that, in the transition from the Roman Empire, the division of labor ran backwards. Instead of more and more specialization, there was less and less. The result was that more and more functions were packed into the vassalage relation. The lord was fighter, military leader, law giver, judge and jury, administrator of the castle, owner of the manor, trainer in arms, economic planner and so on. And these functions were carried out increasingly by illiterate aristocrats. Begin with the political and economic structures of the Roman Empire and subtract the state and urban economy. The residue was the feudal manor.

Vassalage

Under feudalism, there was no centralized state to monopolize the means of violence. Therefore, a social world of exchange was not produced. Instead of exchange, feudalism was built on vassalage. Vassalage is a multicomponent social relation that connects two social actors: a 'man' subordinated to a 'lord'. As Bloch explains:

> Imagine two men face to face: one wishing to serve, the other willing or anxious to be served. The former puts his hands together and places them, thus joined, between the hands of the other man—a plain symbol of submission...the ceremony was called homage...the subordinate was often simply called the 'man' of this lord... In this form the rite bore no Christian imprint. (1961a, pp. 145–146)

According to multiple authors, vassalage was not grounded on traditional structures such as family or lineage (Bloch, 1961a, b; Stephenson, 1942; Ganshof, 1961). Instead, both vassal and lord entered the vassalage relation as individuals. As Ganshof explains, feudalism "pushed to the extremes of the element of personal dependence" (1961, p. xv).

Since Weber (1918/1968) formulated his three ideal types, traditional, legal-rational and charismatic, it has been assumed in sociology that all traditional societies were filled with dense networks of social relations that were grounded in the extended family and governed by habits passed down from ancient times. The personality type fostered by traditional structures was the opposite of an individualist.

While Marx found the source of individualism in the rise of capitalism, historians of feudalism disagree. The highly individualized relation of vassalage occurred more than a millennium before the individualizing effects of capitalism.

While feudalism's vassalage relation was built on individualism, feudalism alone did not produce Europe's massive movement to individualism. The Christian Church was central to its production. Beginning at or around the fourth century CE, the church proscribed an array of practices that had supported tradition and the extended family, while propagating rulings favoring the elemental (nuclear) family (Goody, 1983, p. 56ff). Those rulings included forbidding widows to remarry, while also forbidding both adoption and cousin marriage. Goody asks,

> Why should the Christian Church institute a whole set of new patterns of behavior in the sphere of kinship and marriage, when these ran contrary to the customs of the inhabitants they had come to convert, contrary to the Roman heritage upon which they drew, and contrary to the teachings of their sacred texts? (1983, p. 42)

For Goody the answer lies in how inheritance enriched the church. Inheritance is a covert power structure. It is 'hidden' in that its effects are not immediate, but, instead, work quietly over a substantial time span. According to Goody, the initial conditions for individualism under feudalism include that 20% of couples had only girls and 20% had no children at all. As a result, 40% of marriages had no male heir (Goody, 1983, p. 44). And under Salic law, furthered by the church, only males could inherit. Furthermore, "the celibacy of the priesthood kept wealth from being dispersed into lay hands" (Goody, 1983, p. 81).

The result was an "enormous shift of property…to the hands of the Church" (1983, p. 46). Property was no longer buried with the dead as it had been in virtually all ancient societies. Instead, property was given to the church. As Goody quotes Salvian on page v, "It is hard to leave but little to sons and near ones; it is much worse to be tortured throughout eternity". The church inherited from believers because it had instilled a fear of hell from which salvation was only through the church. Through inheritance the church became the largest landowner in Europe. For example, immediately after the Norman conquest, the income of the church was higher than that of the Crown (Goody, 1983, p. 27). At the same time, "the clan systems of the Ancient Mediterranean … gradually disappeared" (Goody, 1983, p. 233). Vassalage occurred within this highly individualized society.

What are the many components that together form vassalage? At its core is patronage, an important relation stemming from Mediterranean antiquity. Linking a high-status patron to a lower status client, patronage is a mutually agreed upon relation in which the patron provides protection to the client while the client provides services to the patron. Beyond protection, the patron acts as gatekeeper opening or closing the client's access to benefits external to the patron–client relation. Elementary Theoretic experiments show that clients pay fees to patrons for access the patron provides and that the fee varies with the value of access sought (Corra & Willer, 2002).

Beyond clientage, vassalage was the social relation into which was packed many, if not all, of the functions that had been occupational specializations under Imperial Rome. As the division of labor ran backwards, the knowledge base for the exercise of each function became more and more shallow. The lord held court but had no training in law. The lord administrated the fief but had no training in administration. By contrast, the one specialization where skill was maintained at a high level was the ability to fight from horseback.

Taken together, the vassal and his fief composed a small state, a "statelet" in the sense that the functions of a state are carried out within the fief. The fief-holding vassal holds court, administers justice, and has coercive powers to support that administration. That same coercive power makes it possible for him to collect taxes and various fees for usage. For example, the peasant pays taxes to the lord and is required to use, for a fee, the lord's mill for grinding his wheat. As a part of his infeudation, the vassal is obliged to ride to the defense of his lord bringing with him his vassals if he has such. While the fief is a unit of feudal political structure, feudalism is also a "stage in economic history" (Stephenson, 1942, p. 14).

Though a vassal could hold a fief, this does not mean that the vassal 'owned' the fief. It is a normal condition of contemporary society that a thing is owned when the owner exercises exclusive rights to use that thing and benefits from that use (Willer & Emanuelson, 2022). Under feudalism, exclusive rights were unusual. Instead, rights were distributed across many actors from the vassal up the feudal hierarchy to end at some point, at times as high as the king.

An Overview Status and Power

Localization of power was an essential quality of European feudalism and of other social forms called 'feudal'. A lord may have multiple vassalages, and, in turn, the vassal may be lord to multiple men. The lord who institutes multiple vassal relations is subinfeudating, dividing his land into multiple fiefs. The length of a chain of vassalage relations and the number of its branches are ultimately governed by the minimum value needed by the lowest vassal. Beyond day-to-day expenses, the income from the fief must support the lowest vassal's means of violence, including at least two horses and knightly armor. Because a lord may promise mutually contradictory protections, as feudalism developed, the specific conditions of protection promised might come to be prioritized. While the vassalage relation was understood to hold only for the lifetime of lord and man, those positions soon came to be inherited and, with inheritance, greater autonomy for the vassal. Nevertheless, the vassalage relation could be ended by either man or lord. It seems that normally, the reason given for ending the vassalage relation was that the conditions of the agreement had not been fulfilled by the other.

Beneath the high-status oligarchy of lords was a flattened out, low status group of peasants, the majority of whom worked the land and were bound to it. Being bound to the land, peasants were not free to sell their labor to the lord of another fief-holder or to freely travel from the place to which they were tied. All useful work was done by these peasants. Peasants were distinct from slaves by rights. For example, serfs could not be bought and sold. Unfortunately for the peasant, his rights were adjudicated in the court of his lord whose own interests could be expected to dominate his decisions.

Parallel to but quite distinct from the vassalage social structure was the medieval church, the organization of which was not that of lord and vassal. Instead, the offices of the church formed a semi-bureaucratized hierarchy. Unlike members of the feudal status grouping who were frequently illiterate, the officials of the church were, with few exceptions, literate. The power of the church extended from Rome across the whole of western Europe. Power in feudalism was, by contrast, highly localized. While the power of kings covered much broader territories than that of any of their vassals, the church was 'international'. Furthermore, the church came to be much wealthier than any king.

Two Contrasting Medieval Social Structures

In the century since Weber offered his formulations for bureaucratic and traditional organizations, little has been added insofar as theory of those structures is concerned. From his formulations comes a recognition that bureaucracy is more effective at organizing domination than traditional structures. In bureaucracy the structure of domination spreads vertically down the hierarchy through multiple levels while spreading horizontally at the junction of each pair of levels. Today the ability to extend power both vertically and horizontally through multiple relations makes it possible for international corporations to set policies affecting employees across the globe and for nation-states to promulgate laws that apply to many millions of citizens. In medieval Europe, only the church had the ability to extend power both vertically and horizontally. The traditional organizations of feudalism notably lacked those capabilities.

Though Weber attributed obedience in organizations to legitimacy, he said nothing about how beliefs in legitimacy are structured. Nor is anything said about why subordinates would come to believe superiors' claims of right to rule. Instead, Weber offered lists of "characteristics" that are, as will be seen, conditions under which the organization functions. For legal-rational legitimacy, he offered fully six lists concerning conditions in bureaucracies. Here is one such list abstracted from *Part One* of *Economy and Society*.

Weber's Characteristics of Bureaucratic Structure:

> 1) The organization of offices follows the principle of hierarchy: that is, each lower office is under the control and supervision of a higher one. (p. 218)
> 2 (T)he administrative staff should be completely separated from ownership of the means of administration. (pp. 218–219)
> 3 Candidates are selected on the basis of technical qualifications. (p. 220)
> 1) The salary scale is graded according to rank in the hierarchy. (p. 220)
> 2) There is a system of "promotion" according to seniority or to achievement, or both.
> 4) Promotion is dependent on the judgement of superiors. (p. 220)

In fact, the medieval church had many, if not all of these characteristics, while feudal organizations had none.

A theoretical tool is needed to cut through Weber's lists to uncover why some organizational structures, such as bureaucracy, produce domination/obedience and why others, such as feudal structures, do not. To find why we turn again to the Theory Toolbox and pull out Elementary Theory's formulations for strong and null structures. If it can be shown that bureaucratic organizations are strong structures whereas feudal organizations are neutral structures, an important link between structure and domination will have been discovered. Emulating Weber, let's begin with bureaucracy.

Figure 9.5 is the diagram for the simplest organizational *component* consisting of a superior and its immediate subordinates. Organizational structures are built by connecting multiple components both vertically and horizontally. Let's now apply Weber's characteristics as conditions under which the component functions. Applying Weber's first characteristic, the organization of offices follows the principle of hierarchy. Further, applying Weber's second characteristic, all office holders, superiors and subordinates alike, do not own their positions. Now the fourth and fifth characteristic can be applied. By the fourth the salaries of superior and subordinate are ranked and by the fifth there is a system of promotion as judged by the superior. Given these conditions, *the subordinates will*

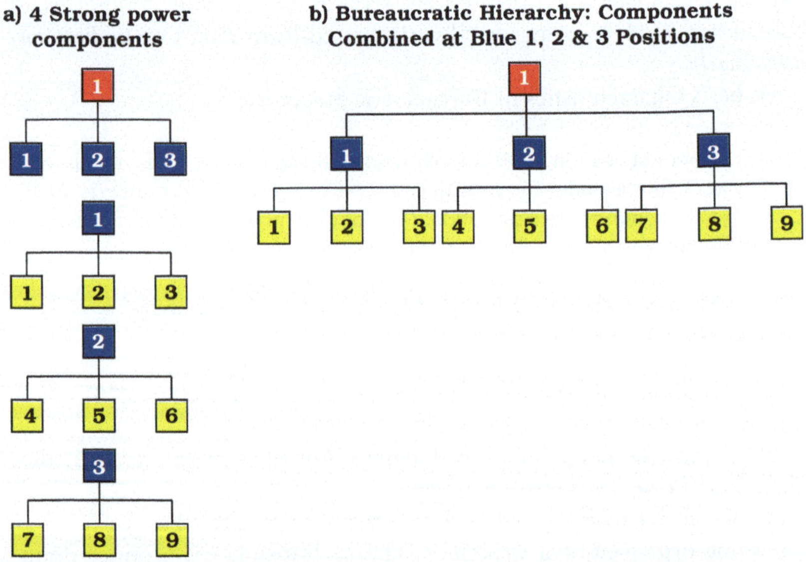

Fig. 9.5 Assembled component power structures to build a model of bureaucracy

compete for promotion. Given that competition, the relations that link superior to subordinates are relations of domination and obedience.

Bureaucratic components are knit together by every position (but for the one at the top and ones at the very bottom) acting as *both* superior and subordinate. Furthermore, the strong structural condition is repeated between each pair of levels. As a result, the competition that produces domination and obedience extends throughout the structure.

Of Weber's five characteristics, 1, 2, 4 and 5 are conditions necessary for the component to function as a part of a strong structure. Only the third characteristic is concerned with the capability of the office holder to expertly apply knowledge. In Weber's Germany, it may have been that bureaucrats were technically qualified. Today, at least in the U.S., the typical qualification for holding bureaucratic office is a college degree, and that degree does not, by any stretch of the imagination, imply technical qualification. Instead, by long tradition, that degree assures minimal qualifications in literacy and simple mathematical capabilities.

The analysis of feudal structures again begins with the Fig. 9.5 diagram, but now we apply Weber's characteristics for feudal structures. They are:

The Characteristics of Feudal Structure: (1918/1968, pp. 255–266, 1070–1100)

1. Positions are not controlled or supervised by superiors.
2. Occupants own the positions they occupy.
3. When positions are not inherited, candidates are selected by status.
4. Positions are funded by fiefs.
5. There is no system of promotion.

In the vassalage relations of feudalism, all positions, but for those at the very bottom and top, act as both superior and subordinate. Nevertheless, feudalism is by no means tight knit. Instead, there is much autonomy between levels. While there are mutually agreed upon rules and obligations connecting lord and man, the relation itself was not supervisory. Regarding points 2 and 3, people come to occupy positions in one of two ways: by inheritance or by matching the status of the position to that of the man. Regarding point 4, there is no central exchequer. Income for positions is not distributed through the structure. Instead, the income of the fief holder is produced on the fief itself. There was no system of promotion because there is no promotion. With no system of promotion, the structure is null. That is to say, the structure itself does not produce power differences.

While the structure does not produce power differences, feudal structures are full of power struggles. It is important to remember that members of the feudal structure are, themselves, members of a military oligarchy. Being armed and dangerous, conflict is endemic. And, like all conflict, the goal is coercion over those who lost the conflict struggle. Furthermore, conflict was ongoing precisely because feudalism is a null structure. Strong structures, like bureaucracies, sort themselves out into stable gradations of power. Null structures do not. The medieval church encouraged the feudal organizations of Europe to do battle in the Middle East, even to sack Constantinople. But it did not encourage power struggles internal to the church itself.

This section began with a question. Why was feudalism the social structure that followed the Roman Empire? The linear-square law assured that slavery must collapse and with that collapse (1) coercion moved from strong to null while (2) privatization was already undercutting the tax base. As Rome's coercive system moved from slave to serf, surpluses squeezed out of the serfs were much smaller than what was squeezed out of slaves. Therefore, the division of labor ran backwards, the money economy failed, the aristocracy moved to the countryside and feudalism was the residue.

Theory Toolbox II

The toolbox employed in this section of the chapter focused on the transitions from strong to null coercive structures. Experimentally grounded theories and models drawn from them located the transition from Empire to feudalism in the transition for strong to null coercive structures. Following Udehn (2001) the models take a "structurally individualistic" form in that social structures are explicitly drawn and are made dynamic by actions of actors whose values and beliefs are acquired from the structures.

Interdependent Political Structures

The Great Wall of China marked a boundary between two distinct ways of life: the settled agricultural society of the Han to the south and the nomadic sheep-raising culture to the north. Despite this division, as Barfield (1989) describes in ***The Perilous Frontier***, the relationship between the Chinese state and the northern horse nomads was not purely adversarial. Instead, it was characterized by a mutual dependence that impacted their political structures. The nomads relied on wealth and goods from China, obtained through trade or tribute, while the Chinese state used the nomads as both a buffer and a tool to maintain stability. This interdependence shaped strategic adaptations on both sides, producing a dynamic balance reflective of the demands of their respective situations.

In the sections that follow, models from the Theory Toolbox are applied in a stepwise manner: first to understand the political structures south of the Great Wall, then to those north of the Great Wall, and finally to explain the dynamics between these two interconnected polities. While Barfield's (1989) analysis offers valuable insights, what additional understanding can we gain from the Theory Toolbox with its models and network diagrams? In particular, can it offer further insight into the underlying mechanisms that maintained their interdependence?

South of the Great Wall

The governance of the Han dynasty can be understood through the Hierarchy-Mobility and Power-at-a-Distance models, which, as previously discussed, show how actors in hierarchical positions exercise power. These models explain how the emperor centralized power within the state's administration, and together they indicate the extent and limits of his reach. Rather than reiterating these models in detail, we focus here on their application to the Han, highlighting the unique dynamics and challenges faced by the Han dynasty. Later, we will extend the previous examination of coercive structures to the Han-Xiongnu interactions and provide a concrete example of how theoretical models apply to the China-Northern Nomad frontier.

First, we apply the Hierarchy-Mobility model. As previously noted, power is concentrated at the top when a bureaucratic system includes mobility and merit-based promotions. During the Han Dynasty, officials were appointed directly by the imperial court, with mid-level officials evaluating their subordinates based on competency, obedience and adherence to laws, while upper-level supervisors used these evaluations to decide on promotions, demotions or dismissals (Twitchett & Loewe, 1986, p. 52). This system allowed the emperor to maintain bureaucratic authority over a wide territory.

The centralization of power led to a concentration of wealth and resources at the top, enabling the emperor to manage the internal stability of the state's administration and external relations. The Han bureaucracy extracted resources from its peasants and funneled them to the top. One specific example of the Han's taxation capabilities was the land tax, which was initially fixed at one-fifteenth of the harvest around 200 B.C. and later reduced to one-thirtieth, remaining unchanged for several centuries (Twitchett & Loewe, 1986). This organization of the taxation system allowed the Han state to amass significant wealth, ensuring a steady accumulation of resources for the state.

The taxation system played a critical role in enabling the Han Dynasty to pursue extensive policies like the Ho Chin policy. The enactment of the Ho Chin policy, as referenced by Barfield (1989), illustrated the immense wealth amassed by the imperial government, which enabled it to appease the nomads north of the wall with substantial quantities of goods. The policy involved paying the Xiongnu's leader north of the great wall vast sums of silk, grain and other valuables to prevent raids and secure peace along the frontier. At times, the Chinese state delivered tens of thousands of rolls of silk and hundreds of thousands of bushels of grain annually, underscoring the immense resources the Han could mobilize (Barfield, 1989).

Centralized power allowed the Han dynasty to manage relations with the Xiongnu by utilizing resource accumulation and strategic tribute payments. However, there were limitations to relying on wealth alone to maintain long-term peace. While the sheer size of the tribute payments was often burdensome to the Han emperor, maintaining an external threat like the Xiongnu served a strategic purpose. As shown below, the presence of a persistent external threat provided the emperor with a justification to maintain strict control internally, suppress potential rebellions and consolidate bureaucratic authority. Though the underlying threat from the Xiongnu persisted, it also worked to the emperor's benefit by reinforcing the need for internal stability.

Consistent with the Power-at-a-Distance model, the emperor extended his power throughout the empire by relying on power relations between pairs of level of the bureaucracy, with appointed officials acting as intermediaries to implement his directives wherever the bureaucracy extended. The Han Emperor's power extended only as far as the bureaucracy reached, allowing for governance only within those boundaries. Still, the

Han administration was extensive, involving multiple bureaus that managed various administrative functions across the empire.

During the Later Han, the secretariat was divided into six bureaus, each responsible for tasks such as handling correspondence with ministers and provincial officials, overseeing foreign relations, managing memorials to the throne and maintaining civil order (Twitchett & Loewe, 1986). Additionally, each commandery was managed by a governor and organized into different bureaus that handled population registration, agriculture, tax collection, military conscription and suppression of banditry. This multi-tiered bureaucratic system illustrates the organizational depth required to manage such a large territory.

Given the size and complexity of the bureaucracy, the extent of the Han emperor's reach was substantial. The success of major public works projects, such as the extension of the Great Wall to defend against northern threats, the construction of extensive canal systems to facilitate transportation, and the expansion of road networks to link distant regions, demonstrates the strengths of the bureaucracy at extending power-at-a-distance to achieve large-scale infrastructure goals. These projects were critical in consolidating central authority and promoting economic integration within the empire.

Comparing the administrative reach of the Ch'in and Han emperors demonstrates how the Power-at-a-Distance model applies to both dynasties, highlighting the differences in the extent and effectiveness of their centralized control. During the Ch'in dynasty, the centralized administrative reforms allowed the First Emperor to exert direct control across a vast territory, eliminating local intermediaries and imposing standardized laws and taxes throughout the empire (Twitchett & Loewe 1986, p. 52). One reform established mutual responsibility groups, as described in Shang Yang's seminal text on legalist reforms. These groups ensured that families were collectively accountable for each other's behavior, embedding a strict hierarchy at the local and familial levels and extending centralized control throughout every stratum of society.

In contrast, the Han dynasty had to accommodate powerful regional lineages that exerted substantial influence in their localities, limiting the reach of the emperor's power to areas where the central bureaucracy was effective. This comparison between the Ch'in's totalitarian centralized governance and the Han's reliance on intermediaries at the local levels aligns with the Power-at-a-Distance model's assertion that bureaucratic power can be maximally exerted at each level of a hierarchical bureaucracy but stops where the reach of the bureaucracy ends.

The reliance on local elites underscores the inherent limitations of centralized power in overcoming entrenched regional status-based influence. This tension between formal bureaucratic authority and informal, status-based influence required a nuanced governance approach, where effective cooperation with local elites was crucial for maintaining overall stability. For instance, during the Han Dynasty, high-ranking landowning families often resisted imperial attempts to restrict private wealth accumulation and manipulated tax laws in their favor, leveraging their lineage-based influence to counterbalance the emperor's directives. This resistance was particularly strong in regions where the state's bureaucracy was weak, such as the older settlements that had ties to high-status lineages. The emperor's efforts to control these areas were frequently undermined by the entrenched influence of local elites (Twitchett & Loewe, 1986). Han emperors sought to suppress the effect of status-influence by moving officials. According to Weber (1901/1951), local appointments were of short duration, "formally of three years after which the official was to be removed to a different province" (p. 48).

Figure 9.6 displays the two hierarchical structures: the power hierarchy on the left and the influence hierarchy on the right. The power hierarchy represents the state's bureaucratic system, where officials (Os) hold positions that are not fixed, allowing for mobility based on performance and loyalty. At the top is the leader (L), who exercises power-at-a-distance, extending authority throughout the bureaucratic structure. On the right, the influence hierarchy is modeled using Hage and Harary's Status Lineage model for structure and Melamed and Savage's Status Factions model for dynamics. Here, the senior lineage elder (E) occupies the topmost position, wielding influence over all other members of the lineage (Fs). Unlike the power hierarchy, this influence is derived from social recognition and lineage prestige rather than formal bureaucratic authority.

In this compound structure, L and E each possess the ability to impose negative sanctions. E's influence is sufficient to organize rebellion or resistance, thereby posing a potential threat to L. Conversely, L can leverage state resources to pay loyal fighters and secure obedience, creating a counter-threat to E. The relationship between L and E is thus characterized by a balance of sanctions, with each actor able to challenge the other.

The power hierarchy's centralization allows L to accumulate greater resources than E, who is constrained by the obligations and limitations

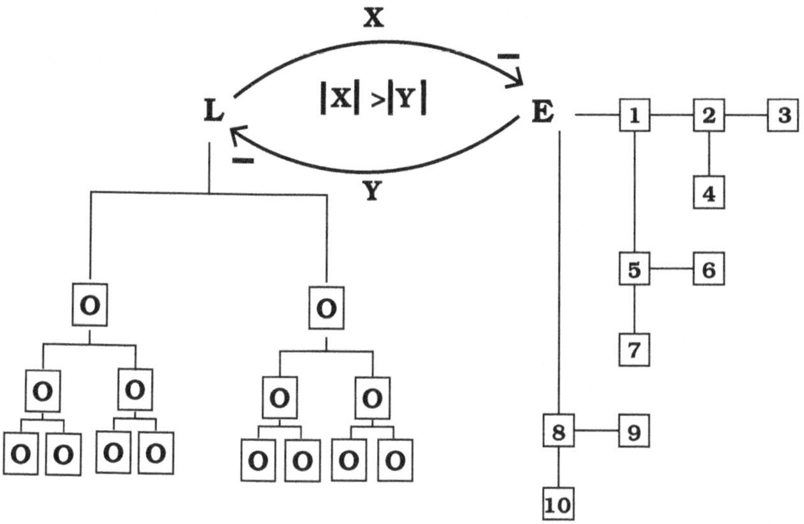

Fig. 9.6 Competing power and status hierarchies

inherent in lineage-based influence. High-status elders like E must distribute resources among their followers, which limits their ability to accumulate wealth. On the other hand, L, operating within the power structure, faces no such restrictions, allowing for greater resource accumulation. According to Elementary Theory's conflict model, victory in a confrontation depends on the ability to impose greater losses on the opposing party. In this context, L's superior resources and broader reach offer an advantage in one-on-one confrontations with E.

The coexistence of these structures underscores the duality of Han governance, where centralized and decentralized forces were in constant interplay. The alliances among noble families presented ongoing challenges to imperial authority, necessitating a dynamic strategy that balanced central power with regional accommodation. Moving from the centralized power dynamics of the Han Dynasty, which relied on bureaucracy and coercion, to the decentralized influence structure of the Xiongnu confederation, we see a marked contrast in how control was achieved.

North of the Great Wall

The Xiongnu nomadic confederation north of the Great Wall had fundamentally different mechanisms for managing influence compared to the Han state. Unlike the Han's agricultural-based society that extended control through a hierarchical bureaucracy, the Xiongnu society relied on a complex system of status lineages with influence through kinship ties and alliances rather than centralized administration. These lineages operated with a high degree of independence, not subordinated to any centralized bureaucratic authority. The leader of the Xiongnu, known as the Shan-yu, occupied the highest position within the elite lineages, deriving his influence from lineage rank and strategic alliances. However, unlike the Han Emperor, whose authority was underpinned by an extensive bureaucratic apparatus, the Shan-yu, not being bureaucratic, faced significant challenges in extending his influence across the different noble lineages of the confederation.

Figure 9.7 depicts two status-lineage structures adapted to the nomadic context: the elite lineage above the dotted line, consisting of families directly related to the Shan-yu, and the subordinate lineage below the dotted line, comprised of a non-elite noble family. The Status-Factions model helps explain the dynamics both within and between these lineages. As shown by the arrow pointed to the right labeled faction effect and status-influence effect, local lineage leaders held significant influence over their members, as their rank gave them influence over all lower-ranked individuals within their lineage, effectively making their faction the largest. The arrow pointing down labeled *only* status-influence effect shows that the Shan-yu's influence extended across multiple lineages due to his position at the top of the highest-ranked elite lineage. However, local leaders, advantaged by both status-influence and faction effects, retained direct control within their own lineages, often surpassing the Shan-yu's influence when it came to their own followers. This dynamic required the Shan-yu to negotiate his influence continuously with the subordinate lineage heads.

Both the Han Emperor and the Shan-yu faced challenges from powerful nobles, though the structural causes of these challenges differed. While the Han Emperor struggled to extend bureaucratic power into local status lineages, the Shan-yu faced difficulties in uniting distinct subordinate lineages under his influence, often requiring negotiation with lineage heads who wielded considerable autonomy. Thus, the Shan-yu's influence was limited and continuously negotiated within the confederation.

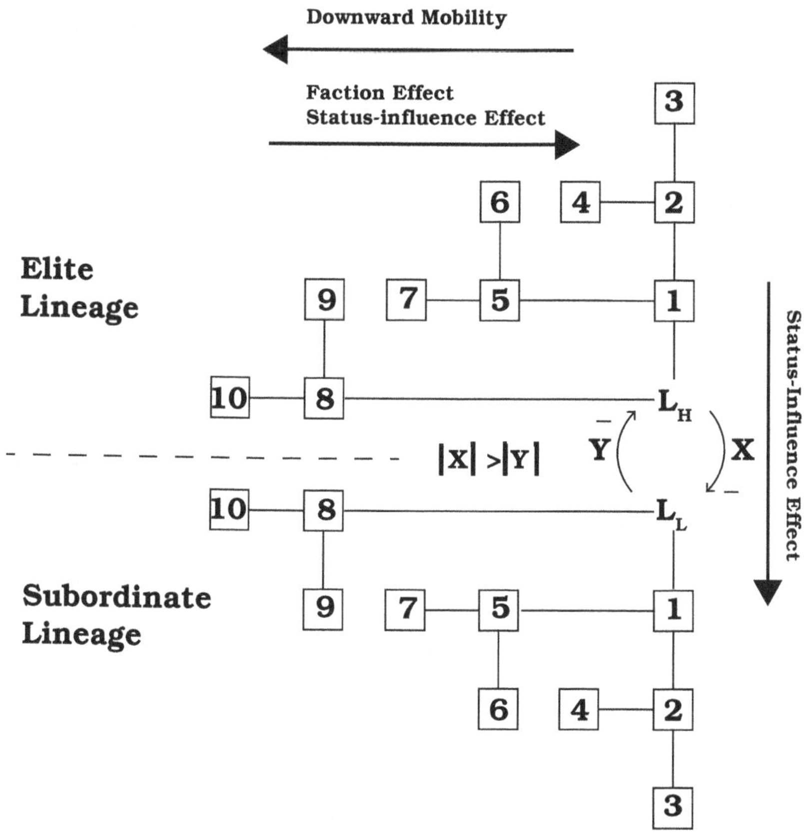

Fig. 9.7 Connected Senior and Cadet Lineages

The nomadic structure presented unique challenges and opportunities for leadership, adding a new dimension to the theme of extending control in a decentralized society. An important limitation of the Shan-yu's influence was the mobility of the horse nomads, which made direct coercion impractical. As Barfield (1989) notes, the nomadic tribes could evade attempts at coercion by moving away from the threat, limiting the effectiveness of centralized influence against them (Barfield, 1989, p. 46).

The Shan-yu needed to balance maintaining influence over the confederation with accommodating the autonomy of noble families. Unable to resort to coercion or bureaucracy, the Shan-yu's control was shaped largely by familial loyalty and strategic alliances among the clans. Both the Han and Xiongnu leaders had to navigate elite interests to ensure stability, but their distinct approaches—bureaucracy versus kinship alliances—illustrate the unique political dynamics that each faced.

Models of Interdependence: Explaining Han-Xiongnu Dynamics

This section examines how the challenges identified in the previous sections, 'South of the Wall' and 'North of the Wall,' were addressed by the Han and Xiongnu leaders. The earlier analysis highlighted the difficulties of maintaining internal stability and managing elite influence. Here, we demonstrate how both the Han emperor and the Xiongnu Shan-yu addressed these challenges by leveraging coercive and gatekeeping strategies in their inter-polity relations. For the Han emperor, indirect coercion was essential to mitigate the influence of the landed gentry by using the Xiongnu threat to centralize authority. For the Shan-yu, his strategies involved extorting tribute from the Han, which enabled him to reward loyal followers and maintain an opulent court, as well as acting as a gatekeeper in trade negotiations with China, thereby controlling access for the tribes.

The Han Emperor's Use of Indirect Coercion. In indirect coercion the coercer offers protection instead of directly threatening the coerced. This concept finds particular relevance here as we explore how the Han Emperor leveraged external threats to centralize his power internally. In the context of Han-Xiongnu relations, indirect coercion played a crucial role in stabilizing internal power dynamics of China by leveraging external threats. To understand the dynamics of indirect coercion, it is essential to examine the interests of both the Han emperor, as the coercer, and the landed nobility, as the coerced. In the absence of external threats, these interests conflict: coercers seek to take as much as possible, the coerced to send as little. Protection from an external threat, in contrast, is a benefit for which the coerced will pay. This section will examine the dynamic between the Han emperor, the Xiongnu raiders and the powerful Chinese status-lineages.

By presenting the nobility with a choice between enduring the risk of nomadic incursions or aligning with imperial authority for protection, the emperor effectively consolidated his centralized power, leveraging indirect coercive structures to align elite interests with those of the central state. This strategy not only provided the nobility with security but also reinforced the emperor's authority by making their loyalty contingent on his ability to protect them from external threats. In doing so, the emperor reduced the likelihood of resistance from the nobility, as their survival and influence increasingly depended on the centralized state. This alignment of interests ultimately strengthened the internal stability of the Han polity by subordinating elite influence to imperial control, reflecting the emergent properties of a coercive, centralized governance structure.

Direct Coercion and Gatekeeping: The Shan-yu's Strategies. While the Han emperor relied on indirect coercion to control the nobility, the Xiongnu Shan-yu leveraged direct coercion to exploit the vulnerabilities of the Han state. Direct coercion depends on the strength of negative sanctions; if these sanctions are severe and costly, compliance is more likely as the coerced party will seek to avoid the sanction. The coercer must also be capable of effectively deploying these sanctions to ensure compliance.

The Shan-yu exercised direct coercion through the mobility of his cavalry and the significant threat it posed to Chinese interests. The Xiongnu cavalry could strike deep into Chinese territory, causing substantial damage to the estates of the landed nobility. This capacity for rapid and damaging raids made Han retaliation ineffective, as the Chinese army lacked the mobility to match the Xiongnu. As Barfield (1989) notes, "The Xiongnu imperial government owed its financial and political stability to its exploitation of resources outside the steppe" (pp. 36–37). The Shan-yu used this capability to directly coerce the Han state, extracting tribute and consolidating his authority by demonstrating his ability to impose costs if his demands were not met.

The tribute extracted through the Ho-chin policy provided the Shan-yu with resources to support his court and reward loyal noble lineages (Barfield, 1989). Redistribution, as described by Sahlins (1963), involves leaders accumulating resources from community members and redistributing them to reinforce loyalty and consolidate their authority. However, the nomadic lifestyle of the Xiongnu did not produce the surplus necessary to support lavish redistribution on its own. Unlike the settled, agricultural and bureaucratic state of China, which generated surplus resources through systematic taxation, the Xiongnu relied heavily on tribute from China. This redistribution of tribute allowed the Shan-yu to reward loyalty and maintain stability within the confederation. However, the limited and inconsistent nature of tribute meant that it was not always enough to ensure loyalty across all tribal leaders, necessitating additional strategies to maintain cohesion.

To address these limitations, the Shan-yu employed gatekeeping as an additional strategy to maintain authority. In gatekeeping the gate keeper controls access to valued goods or services, establishing a power relationship where clients must pay for access (Corra & Willer, 2002). This strategy is effective when the gatekeeper monopolizes access and imposes penalties for unauthorized access.

The Shan-yu acted as a gatekeeper, selectively granting access to Chinese border markets. According to Barfield (1989, p. 48), "Access to Chinese border markets was crucial for the Xiongnu tribes, as unsold meat and animal products before winter could lead to substantial losses and reduced resources to survive the cold". By controlling this access, the Shan-yu could reward loyal tribes while those denied access faced significant hardships during winter, effectively consolidating his influence. To strengthen his position, the Shan-yu guarded his monopoly over trade negotiations with China (Barfield, 1989). He minimized unauthorized border crossings, as corrupt border guards could undermine his monopoly and weaken his leverage over tribal leaders.

Ultimately, while the Shan-yu's gatekeeping strategy relied on controlling access to external resources to reinforce loyalty, the Han emperor's indirect coercion leveraged the Xiongnu threat to maintain control internally. Both approaches played pivotal roles in stabilizing their respective polities, each adapting to the unique challenges posed by their political environments.

Stabilizing Relations Between Han and Xiongnu

The boundary between the Chinese state and the nomadic Xiongnu remained stable for centuries through the interplay between the Han and Xiongnu policies. Indirect coercion and direct coercion served to balance internal and external pressures, maintaining a stable but complex equilibrium along the frontier. The Han emperor used the threat of Xiongnu incursions to indirectly coerce the landed gentry, thereby aligning their interests with those of the central state. In turn, the Xiongnu Shan-yu rewarded loyal followers through the redistribution of resources obtained from China and through gatekeeping controlling access to the valuable Chinese markets.

Theory Toolbox III

In this section, the theory toolbox application focused on the strategic mechanisms the Han emperor and the Xiongnu Shan-yu employed to stabilize their respective polities through inter-polity relations. The models utilized—indirect coercion, direct coercion, and gatekeeping—explained how each leader addressed internal challenges by leveraging external threats and opportunities.

The indirect coercion model was instrumental in explaining how the Han emperor mitigated elite influence. By offering protection from the Xiongnu raids, the emperor was able to align the interests of the nobility with those of the central state, thus reducing internal resistance and reinforcing centralized authority. This model demonstrated how external threats could be strategically employed to consolidate internal power, ensuring that elite status influence did not destabilize the bureaucracy.

For the Xiongnu Shan-yu, direct coercion and gatekeeping explain how external tribute from the Han was used to maintain loyalty and stability within the confederation. The Shan-yu directly coerced China into providing tribute, which he redistributed among his followers to strengthen alliances. However, since the tribute was insufficient to ensure loyalty across all tribes, the Shan-yu also acted as a gatekeeper, controlling access to Chinese markets and selectively rewarding tribes that demonstrated loyalty. This combination of models illustrated how strategic control over both resources and opportunities for trade helped the Shan-yu maintain authority over the nomadic tribes.

References

Aberle, D. (1961). Matrilineal descent in a cross-cultural perspective. In D. Schneider & K. Gough (Eds.), *Matrilineal kinship* (pp. 655–730). University of California Press.

Anderson, D. (1994). *The Savannah River chiefdoms.* University of Alabama Press.

Barclay, P., & Benard, S. (2013). Who cries wolf, and when? Manipulation of perceived threats to preserve rank in cooperative groups. *PLoS ONE, 8*(9), e73863.

Barfield, T. J. (1989). *The perilous Frontier: Nomadic empires and China.* Blackwell Publisher.

Barfield, T. J. (1990). Tribe and state relations: The inner Asian perspective. In P. Khoury & J. Kostiner (Eds.), *Tribes and state formation in the middle east* (pp. 153–184). University of California Press.

Benard, S. (2012). Cohesion from conflict: Does intergroup conflict motivate intragroup norm enforcement and support for centralized leadership? *Social Psychology Quarterly, 75*(2), 107–130.
Benard, S., & Doan, L. (2011). The conflict-cohesion hypothesis: Past, present and possible futures. *Advances in Group Processes, 28*, 189–225.
Bloch, M. (1961a). *Feudal society: The growth of ties of dependence* (L. A. Manyon, Trans.). The University of Chicago Press.
Bloch, M. (1961b). *Feudal society: Social classes and political organization* (L. A. Manyon, Trans.). The University of Chicago Press.
Bornstein, G. (2003). Intergroup conflict: Individual, group and collective interests. *Personality and Social Psychology Review, 7*(2), 139–145.
Bornstein, G., Erev, I., & Rosen, O. (1990). Intergroup competition as a structural solution to social dilemmas. *Social Behavior, 5*, 247–260.
Bornstein, G., Gneezy, U., & Nagel, R. (2002). The effect of intergroup competition on intragroup coordination: An experimental study. *Games and Economic Behavior, 4*, 1–25.
Buccellati, G. (1998). The role of socio-political factors in the emergence of a "public" and "private" domains in early Mesopotamia. In M. Hudson & B. Levine (Eds.), *Privatization in the ancient near east and classical world* (Vol. I, pp. 129–152). Harvard University Press.
Carneiro, R. L. (1970). A theory of the origin of the state. *Science, 169*, 737–738.
Carneiro, R. L. (1981). The chiefdom: Precursor of the state. In G. D. Jones & R. K. Kautz (Eds.), *The transition to statehood in the New World* (pp. 37–79). Cambridge University Press.
Carneiro, R. L. (1991). The nature of the chiefdom as revealed by evidence from the Cauca Valley of Colombia. In A. T. Rambo & K. Gillogly (Eds.), *Profiles in cultural evolution: Papers from a conference in honor of Elman R. Service. Anthropological papers museum of anthropology* (pp. 167–190). University of Michigan.
Carneiro, R. L. (1998). What happened at the flashpoint? In E. M. Redmond (Ed.), *Chiefdoms and chieftaincy in the Americas* (pp. 19–42). University Press of Florida.
Carneiro, R. L. (2012). The circumscription theory: A clarification, amplification, and reformulation. *Social Evolution and History, 11*(2), 5–30.
Corra, M., & Willer, D. (2002). The gatekeeper. *Sociological Theory, 20*, 180–207.
Dandmayev, M. (1998). An age of privatization in ancient Mesopotamia. In M. Hudson & B. Levine (Eds.), *Privatization in the ancient near east and classical world* (Vol. I, pp. 197–222). Harvard University Press.
Earle, T. (1991a). Property rights and the evolution of chiefdoms. In T. Earle (Ed.), *Chiefdoms: Power, economy and ideology, school of American research advanced seminar series* (pp. 71–99). Cambridge University Press.

Earle, T. (1991b). The evolution of chiefdoms. In T. Earle (Ed.), *Chiefdoms: power, economy and ideology, school of American research advanced seminar series* (pp. 1–15). Cambridge University Press.

Edzard, D. O. (1998). Private land ownership and its relation to "God" and the "State" in Sumer and Akkad. In M. Hudson & B. Levine (Eds.), *Privatization in the ancient near east and classical world* (Vol. I, pp. 109–128). Harvard University Press.

Elias, N. (1994). *The civilizing process.* Blackwell.

Emanuelson, P., & Willer, D. (2015). External threat as coercion. *Journal of Social Structure, 16*(6), 1–16. https://www.cmu.edu/joss/content/articles/volindex.html

Emanuelson, P., & Willer, D. (2018). Social structures in transition: Applications of two theories to chiefdoms. *Socius: Sociological Research for a Dynamic World, 4*, 1–14.

Finer, S. E. (1997). *The history of government: The intermediate ages.* Oxford University Press.

Flannery, K. (1999). Process and agency in early state formation. *Cambridge Archaeological Journal, 9*(1), 3–21.

Flannery, K., & Marcus, J. (2000). Formative Mexican chiefdoms and the myth of the "mother" culture. *Journal of Anthropological Archeology, 19*, 1–37.

Flannery, K., & Marcus, J. (2003). The origin of war: New 14C dates from ancient Mexico. *Proceedings of the National Academy of Sciences of the USA, 100*(20), 11801–11805.

Flannery, K., & Marcus, J. (2012). *The creation of inequality: How our prehistoric ancestors set the stage for monarchy, slavery and empire.* Harvard University Press.

Ganshof, F. L. (1961). *Feudalism* (P. Grierson, Trans.). Harper & Row.

Gavrilets, S., & Fortunto, L. (2014). A solution to the collective action problem in between-group conflict and within-group inequality. *Nature Communications, 26*, 1–11.

Gernet, J. (1982). *A history of Chinese civilization* (2nd ed., J. R. Foster & C. Hartman, Trans.). Cambridge University Press.

Goldman, I. (1960). The evolution of Polynesian societies. In S. Diamond (Ed.), *Culture in history: Essays in honor of Paul Radin* (pp. 687–712). Columbia University Press.

Goody, J. (1983). *The development of the family and marriage in Europe.* Cambridge University Press.

Hage, P., & Harary, F. (1996). *Island networks.* Cambridge University Press.

Hansen, M. H. (2006). *Polis: An introduction to the Ancient Greek city-state.* Oxford University Press.

Hedström, P., & Udehn, L. (2009). Analytic sociology and theories of the middle range. In P. Hedström & P. Bearman (Eds.), *The Oxford handbook of analytic sociology* (pp. 24–47). Oxford University Press.

Hudson, M. (1995). *The privatization of land: How it all began*. New York University Colloquium on Land and Antiquity, p. 5.

Hudson, M. (2018). *...and forgive them their debts: Lending, foreclosure and redemption—From bronze age finance to the jubilee year*. ISLET-Verlag.

Hudson, M. (2023). *The collapse of antiquity: Greece and Rome as civilization's oligarchic turning point*. Islet Verlag.

Hudson, M., & Levine, B. (1996). *Privatization in the ancient and classical world* (p. 5). Peabody Museum Bulletin.

Hudson, M., & Levine, B. (Eds.). (1998). *Privatization in the ancient near east and classical world* (Vol. I). Harvard University Press.

Kahneman, D., & Tversky, A. (1979). Prospect theory: An analysis of decision under risk. *Econometrica, 47*(2), 263–292.

Keesing, R. (1975). *Kin groups and social structure*. Holt, Rinehart, and Winston.

Kelly, R. C. (2000). *Warless societies and the origin of war*. The University of Michigan Press.

Kilduff, G. J., Willer, R., & Anderson, C. (2016). Hierarchy and its discontents: Status disagreement leads to withdrawal of contributions and lower group performance. *Organizational Science, 27*(2), 373–390.

Kirch, P. V. (2010). *How chiefs became kings*. University of California Press.

Kurtz, D. V. (2012). State origins: Classifications and reformulations. *Social Evolution and History, 11*, 67–69.

Malo, D. (1903). *Hawaiian antiquities* (N. B. Emerson, Trans.). Hawaiian Gazette Co. Ltd. (Original work published 1898).

Marcus, J. (1998). The peaks and valleys of ancient states. In G. Feinman & J. Marcus (Eds.), *Archaic states* (pp. 59–94). School of American Research Press.

Marcus, J., & Gary, M. F. (1998). Introduction. In G. M. Feinman & J. Marcus (Ed.), *Archaic states* (pp. 3–29). School of American Research Press.

Marx, K. (1970). *A contribution to the critique of political economy* (M. Dobb, Ed. & S. W. Ryazaskaya, Trans.). International Publishers. (Original work published 1859).

Melamed, D., & Savage, S. (2013a). Status, numbers, and influence. *Social Forces, 91*, 1085–1104.

Melamed, D., & Savage, S. (2013b). Predicting social influence with faction sizes and relative status. *Social Science Research, 42*, 1346–1356.

Melamed, D., & Savage, S. V. (2016). Status, faction sizes, and social influence: Testing the theoretical mechanism. *American Journal of Sociology, 122*(1), 201–232. https://doi.org/10.1086/686943

North, D., & Thomas, R. (1973). *The rise of the Western world: A new economic history*. Cambridge University Press.

Parkin, R. (2004). Introduction. In R. Parkin & L. Stone (Eds.), *Kinship and family: An anthropological reader* (pp. 29–42). Blackwell.

Pettit, N., Yong, K., & Spataro, S. (2010). Holding your place: Reactions to the prospect of status gains and losses. *Journal of Experimental Social Psychology, 46*, 396–401.

Popper, K. (1959). *The logic of scientific discovery*. Hutchinson. (Original work published in 1934).

Puurtinen, M., & Mappes, T. (2009). Between-group competition and human cooperation. *Proceedings of the Royal Society, 276*, 355–360.

Redmond, E., & Spencer, C. (2012). Chiefdoms at the threshold: The competitive origins of the primary state. *Journal of Anthropological Archaeology, 31*, 22–37.

Sahlins, M. D. (1958). *Social stratification in Polynesia*. University of Washington Press.

Sahlins, M. (1972 [1963]). Poor man, rich man, big man, chief: Political types in Melanesia and Polynesia. In *Stone age economics* (pp. 281–303). Aldine-Atherton.

Schneider, D. (1961). Introduction: The distinctive features of matrilineal descent groups. In D. Schneider & K. Gough (Eds.), *Matrilineal Kinship* (pp. 1–32). University of California Press.

Simmel, G. (1955). *Conflict & the web of group-affiliations*. (K. H. Wolf & R. Bendix). Free Press.

Simpson, B., Willer, R., & Ridgeway, C. (2012). Status hierarchies, and the organization of collective action. *Sociological Theory, 30*(3), 149–166.

Stephenson, C. (1942). *Mediaeval Feudalism*. Cornell University Press.

Twitchett, D., & Loewe, M. (1986). *The Cambridge history of China: Volume 1: The Ch'in and Han Empires, 221 B.C.–A.D. 220*. Cambridge University Press.

Udehn, L. (2001). *Methodological individualism: Background, history, and meaning*. Routledge.

Van Vugt, M., Hogan, R., & Kaiser, R. (2008). Leadership, followership, and evolution. *American Psychologist, 63*(3), 182–196.

Weber, M. (1951). *The religion of China*. The Free Press. (Original work published 1901).

Weber, M. (1968). *Economy and society*. University of California Press. (Original work published 1918).

Weber, M. (1988). *The agrarian sociology of ancient civilizations*. Verso. (Original work published 1896).

Willer, D. (1984). Analysis and composition as theoretic procedures. *The Journal of Mathematical Sociology, 10*, 241–270.

Willer, D. (1987). *Theory and the experimental investigation of social structures*. Gordon and Breach Science Publishers.

Willer, D. (2003). Power-at-a-distance. *Social Forces, 81*, 1295–1334.

Willer, D., & Emanuelson, P. (2021). Theory and the replication problem. *Sociological Methodology, 51*(1), 146–165. https://doi.org/10.1177/0081175020955216

Willer, D., & Emanuelson, P. (2022). Theory toolbox for historical explanation: An essay in analytic sociology. *Methodological Innovations, 15*(3), 236–250. https://doi.org/10.1177/20597991221119339

Willer, D., Emanuelson, P., Lovaglia, M., Simpson, B., et al. (2014). Elementary theory: 25 years of expanding scope and increasing precision. *Advances in Group Processes, 31*, 175–217.

Willer, R. (2004). The effects of government-issued terror warnings on presidential approval ratings. *Current Research in Social Psychology, 10*, 1–12.

Willer, R., Youngreen, R., Troyer, L., & Lovaglia, M. J. (2012). How do the powerful attain status?: The roots of legitimate power inequalities. *Managerial Decision Economics, 33*(5–6), 355–367.

Wright, H. T. (1977). Recent research on the origin of the state. *Annual Review of Anthropology, 6*, 379–397.

Wright, H. T. (2006). Early state dynamics as political experiment. *Journal of Anthropological Research, 62*, 305–319.

CHAPTER 10

Conclusions

To rebuild sociology as a theoretical science requires a fundamental change in thinking about research. That change is needed because, in spite of superficial similarities, empiricism and science are exact opposites. For the empiricist ideology of induction, generalizations reflect regularities of the world. Therefore, research should be designed to find regularities. By contrast, scientific theory is general because it imposes its order on the world. Therefore, theory should be structured such that it imposes order on observations. Since they are opposites, doing empiricism leads away from doing science and doing science leads away from doing empiricism. Empiricism's ideology of induction is the intellectual underbrush that must be swept away so that a theoretic science of sociology can be built.

A search through the literature of the successful sciences such as physics, chemistry and biology to find books on research methods will be disappointing. Looking instead through the literature of sociology, social psychology and related fields, a large methodological literature on how to carry out research will be found. Furthermore, there is a sharp line dividing those fields where replication of research results is routine and those fields where it is not. That dividing line is exactly the same as the one separating the successful sciences from the empiricism of psychological and social sciences. A pattern begins to emerge. The methods texts presume that they point the way to science. They do not.

D. Willer, P. Emanuelson, *The Logic of Theoretical Sociology*,
https://doi.org/10.1007/978-3-031-91709-7_10

To make sense of the juxtaposition of an extensive methodological literature and the failure of the methods of that literature to produce science is a task for a history of ideas. Chapter 2 of this book gives something of that history albeit in brief. In that chapter the historical development of the empiricist ideology was traced from Mill, through Pearson to Fisher. That ideology asserts that induction from given particulars to the fully general is the method of the sciences. Strangely absent is the inductive methodology that could produce fully general knowledge from given particulars. In fact, science does not employ induction. Science employs theory.

What is theory and how is it built? Chapter 3 reviews the thinking of four notable scholars: Merton a sociologist, and three philosophers of science, Hempel, Popper and Kuhn. Well known is Merton's proposal for middle range theory. Presumably, middle range theory would avoid the grand schemes of a Parsons and the empiricism of a Lazarsfeld by falling somewhere between. Less well known is Merton's proposal that, once theories of the middle range have been developed, they should be consolidated to gain broader scope. The idea of building broader scope theory block by block by consolidating middle range theories is an attractive image. Unfortunately, Merton says little about how middle range theories should be constructed and how they could be consolidated.

Sadly, the three philosophers of science also say little about how theory is to be built. Given the poorly developed social theory of Merton's era, it is understandable that his examples of theory are not instructive. The three philosophers of science are not subject to that limitation. They have the whole of the literature of science upon which to draw. Nevertheless, little is said by any of the three about how theory is built, out of what components it is constructed, or how theory is used in explanation. Nevertheless, there are useful ideas in Popper's philosophy of science. By contrast, Kuhn muddies the water by offering a description of 'science' in terms that suggest that science is religion. It is not.

It is inevitable that replication would be found to be problematic in fields that adopted Fisher as the design for their experiments. It is inevitable because Fisher-design relies upon induction to generalize research findings beyond the laboratory. But as shown there is no inductive method. Nor can there be for, as Popper has demonstrated, the attempt to base the principle of induction on experience leads to infinite regress. Therefore, the research results of Fisher-design experiments are not general. They are particular to the time and place where they were observed. Since results are particular, not general, there is no reason why Fisher-design experiments should replicate. In fact, a substantial proportion do not.

Nevertheless, as was shown in Chap. 4, adding theory to a Fisher-design produces a hybrid design and hybrid designs do replicate. That hybrid designs replicate is due, not to induction, but to the addition of theory as a part of the design. Looking back at Chap. 2 it will be remembered that Fisher spent many years at the Rothamsted research labs at a time when the science of biology in general and the theory of genetics in particular were rapidly developing. Is it possible that Fisher, when at Rothamsted, was working with hybrid designs that owed their generality to theory and not as he believed to induction?

The way that a worldview conceptualizes the phenomena of a theory is an integral part of how theory is developed. The discussion of worldviews in Chap. 5 began with examples from the physics of motion. Aristotle's view was that motion must always be the movement of a body through something—presumably that something was air or water. For Aristotle movement through nothingness, through empty space, was not just impossible. It was wrong-headed to think that motion through nothing is possible. Galileo's great innovation was to *negate* Aristotle's worldview by reconceptualizing motion as moving through nothing—through empty space. Here is something important we learn from Galileo. Negating a commonly held worldview is a good way to start the development of a new worldview.

Treating motion as through the nothingness of geometric space, as did Galileo and later Newton, radically simplified that mathematics of motion. It also simplified motion as observed. Remembering that Galileo had no accurate way to measure short time intervals, he found a happy shortcut to falsifying Aristotle by dropping cannon balls and musket balls from the tower of Pisa. But Chap. 5 left negation as a way to develop new worldviews incomplete. Just as Galileo negated Aristotle, Einstein negated the Galilean—Newtonian synthesis of space as a nothingness. For Einstein, space is a thing with knowable properties. Space has a maximum measurable speed—the speed of light. Space is a thing through which electromagnetic waves and gravity waves are propagated. Space has a shape that is bent by gravity.

The place of negation in generating worldviews in sociology is no less significant than in physics though, at times, the worldviews of society seem to run backward from more sophisticated to less. For example, for George Homans there is no such thing as social structure. There are only the actions of men. Remove people from society and there is no residue. Contrast Homans' empty social world with Marx's that is full of social relations. As Marx was quoted earlier in this book, "In the social production of their existence, men inevitably enter into definite relations which are independent of their will". That Marx's worldview is deterministic is quite often seen as a fault. To the contrary, Marx's determinism and the determinism of any social theory is a virtue. Indeed, any theory that sees action in society as indeterminate will be difficult if not impossible to test.

Why is there distrust of determinism in social theory? Perhaps deterministic theories necessarily treat humans as automatons devoid of free will. The humans of deterministic theory are like Pavlov's dog. When the bell rings, the dog salivates. When the bell rings we begin working on the assembly line. But it is wrong to see deterministic theories as necessarily treating humans as automatons. Only badly constructed deterministic theories do. No economist, no game theorist and no sociological structuralist sees the pursuit of their interest as a mechanical reaction to given conditions. And no economist, game theorist or no sociological structuralist sees a failure to recognize and act in pursuit of an interest as an expression of the freedom of the human spirit. Freedom and stupidity are not the same thing.

Granting that determinism is a good quality in social theory, what should we make of the pluralism of Weber's sociology? Is his sociological theory indeterminate and thus untestable? First, it is useful to note that there are many scholars in sociology who, in search of explanations, blend Marxian and Weberian theory. That blending represents some of the best thinking in the field. Second, it is helpful to remember that Weber was an historian and legal scholar. He was no logician and certainly not a mathematician. Could Weber's indeterminism be only apparent and not fundamental to his thought? Perhaps this much should be said. When faced with a problem of historical explanation, for example the decay of ancient civilization, Weber (1950) denied that he is offering an explanation. Nevertheless, Weber does just that. He offers a theoretically sophisticated explanation for the fall of the Roman Empire in the west.

Comparing the development of worldviews across sciences one is immediately struck by the development of a geometry of motion by Galileo. His pioneering work was a turning point. The further development of the theory of motion by Newton and subsequent developments of theory in chemistry and biology were each associated with a geometry. Thus far, in the social sciences, only Elementary Theory has been developed with geometric representations. A geometry in economics for economic structures is notable in its absence.

As seen in Chap. 8, Status Characteristics Theory is well into its second generation of development by a number of notable scholars. Why has that theory not developed a geometry for structures of influence? There is a reason. Prior to Melamed and Savage (2013a, 2013b), Status Characteristics Theory was scope limited to dyads. Early in Chap. 9's discussion of chiefdoms, theory was needed to cover influence structures substantially larger than the dyad. A theory was cobbled together by borrowing formulations from network analysis to extend Melamed and Savage. Better would be new formulations for n-person structures of influence, developed with its own geometry.

What is gained by developing a geometry to represent social structures as a component of social theory? First, it is much easier and more precise to draw a diagram of a social structure than to imagine the structure without a representation. Furthermore, it is simply not possible to precisely communicate the exact configuration of a structure without an explicitly drawn representation. Second, the most precise way to develop theory is to begin with the simplest ideas and then build more complex formulations by combining simpler ones. Parallel to that step-by-step process is the step-by-step development of geometric representations of the theory's phenomena. Because the development of a geometric representation is inherently a step-by-step process, that development forces other components of the theory to also be developed by combining simple elements to build more complex formulations.

Sociological theories that lack a means to picture social phenomena also lack a technical vocabulary. In Elementary Theory, a technical vocabulary is introduced as more complex terms are introduced by defining them using simpler terms already introduced. As will be remembered, Weber's theorizing began with social actor then moved to social relations heading toward social structures. However, his theorizing did not produce an internally consistent technical vocabulary because his more complex terms were *not* defined by simpler ones. More specifically, social relations were not defined by paired social actions, and social structures were not defined by social relations connected in various ways. Instead, in Weber's social theory, each term's definition is entirely independent from every other term. By contrast, when theory contains a modeling procedure, mental images develop side by side with the terms of the theory. The result is an internally consistent technical vocabulary. Modeling makes the project of building the vocabulary much simpler and more straightforward than the task attempted by Weber.

Testing a theory so introduced entails an analytic factor that is almost universally overlooked. An example will help explain the point being made. Imagine an experiment designed to test a centralized exclusionary structure. The experimentalist builds the replica for the experiment using the structure of the diagram given in the theory. The values of the modeled actors translate as the payoffs to be gained by experimental subjects while their beliefs are the structure as built and the values and beliefs of the others in the study. These are the initial conditions of the model as realized in the experiment. To be observed are the dynamics of the interacting subjects and the distribution of payoffs at the conclusion of each run.

What could go wrong? Could this design falsify the theory being tested? The interactions expected consist of a series of bids that move the payoffs toward the end of the negotiation set favored by the centrally placed subject. Assume that bidding does not occur. The exchange ratio remains static at a medial rate. Theory says that that can happen only if (1) the peripherals are acting collectively or (2) some or all of the subjects are not motivated as expected by the theory. If the first, further components of the theory that explain collective action can be brought to bear. If the second, the experiment is redesigned to clarify the payoff system.

Is it possible that failure of an experiment does not falsify its theory? It seems that failure does not necessarily falsify. It is not a falsification because the failure is due either to (1) not including all of the social structures of the experiment in the replica or (2) a faulty realization of the model in the lab. In either case, the theory has not been falsified. Indeed, it is difficult to imagine how the experiment could falsify its theory. Nevertheless, the experiment is, without a doubt, a test of the theory. Should there be concern that the experiment is not 100% synthetic—that the theory is becoming analytic? 'Analytic' means 'necessarily true as a consequence of its logical structure'. There is no reason for concern. Whether the theory can or cannot be falsified, it nevertheless effectively explains and predicts and those are the goals of science. Nevertheless, the simultaneous analytic/synthetic quality of theory is by no means limited to the above example. It is common across the sciences. If so, one of the tests of a good theory is its ability to resist falsification.

Given that building from simple to complex is the way that theory should be built, what should the end product of theory building look like? Should one be looking for relations between, for example, wealth and education status? Between place of residence and political persuasion? Certainly not. As already explained, the goal of theory building is not to make generalizations ordered by the phenomenal world. That is empiricism. Science is the reverse. The goal of scientific theory is to develop a well-ordered image of the world that is imposed on data. Here is a useful clue. Empirical generalizations have two terms. Scientific laws have three. The following are examples of laws.

The equation, D = VT does not mean that the measures of distance, velocity and time are approximately equal, but that V is D/T, that velocity *is* distance divided by time. And F = MA *means* that acceleration of any given mass is proportionate to the force applied to it. It also means that, for any given force, mass and acceleration vary inversely… and similarly. In Elementary Theory the rate of sanction flows is given by actors' values and the qualities of the structure in which the sanctions are flowing. Just as V = D/T defines velocity, exchange ratio is defined by actors' values and the qualities of the structure. These are definitional and not subject to falsification.

Measurement in science is not scaling. In scaling, a value is found by summing and averaging subjects' responses to questions. Scaling is generalizing and is ordered by the phenomenal world. That is empiricism. Science is the reverse. In scientific measurement, the dimensions of both sides of F = MA are mass, and distance over time squared. The same holds for Einstein's postulation of $E = MC^2$ where the dimensions on both sides are mass and distance over time squared. Einstein's particular genius was discovering that V, the velocity was the speed of light. Given 'V' the two sides of the equation are identical as they must be for the law is an identity. Since it is an identity, it is true by definition and cannot be arrived at by induction or any other form of generalization and it also cannot be falsified.

Sociological theories are notably lacking in precision in measuring social phenomena. In the successful sciences, measurement is either fundamental or derived. The systematic development of fundamental and derived measures stands in the greatest possible contrast to the approach typical of attitude measurement. For example, the 'f' test sums and averages responses to an array of questions. The result is presumably an index value for 'f', the fascist rigidity of thinking of the respondent. That this 'testing' and averaging uncovers something fundamental to fascist rule is difficult to believe. In any case, it is not good science.

A measure is fundamental when an event is directly observable. For example, in Newtonian dynamics, mass, distance and time are fundamentally measured. The forces and energy of the system are measured by derivation through laws. Similarly, in the power structures modeled by Elementary Theory, the number of resources that flow between and through positions, and the elapsed time of those flows, are directly observable as are the number of positions among which the flow occurs. With the three fundamentally measured, the amount of power exercised is a derived measure: it is the ratio of flows to and from each position.

The ninth chapter of this book is concerned with scientific explanation. In that chapter explanations apply theories that have been tested in experiments. In explanation, experimentally tested theories are pulled from the toolbox and applied to the phenomenon at hand. For example, to explain why chiefdoms are warlike one pulls from the toolbox an experimentally tested theory that relates outside conflict to internal solidarity. To explain the transition from slavery to serfdom as Rome fell, theory of strong and null connected structures is pulled from the toolbox to explain the falling rate of coercive exploitation.

Applications of theory from the toolbox are not like Hempel's "deductive nomological" explanation or Popper's similar formulation. They are not because there are no major or minor premises. To the contrary, the best way to understand scientific explanation using a theory toolbox is to think of it as experimentation run backwards. In experimentation one builds a model, then derives the design of the experiment that is the replica of the model. Unlike experimentation, scientific explanation is not dominated by the power to control the phenomena. To the contrary, the incident to be explained is given and the explanation consists of fitting a known experimentally tested model to that which has been observed.

Having falsified induction as a method of science, there remains only one route to general knowledge. That route is theory. How is theory built? How is theory applied? Across an array of literatures almost nothing has been said. This book replaces that silence with a great many examples of how theory is built, how it is tested and how it is applied for scientific explanation. Begin building a theory with a diagraming method to picture the phenomena. Parallel development of a technical vocabulary together with careful development of fundamental and derived measures make the theory more analytic. The result will be theory that is less falsifiable without inhibiting its explanatory power.

References

Melamed, D., & Savage, S. (2013a). Status, numbers, and influence. *Social Forces, 91*, 1085–1104.

Melamed, D., & Savage, S. (2013b). Predicting social influence with faction sizes and relative status. *Social Science Research, 42*, 1346–1356.

Weber, M. (1950). The social causes of the decay of ancient civilization. *The Journal of General Education, 5*(1), 75–88. http://www.jstor.org/stable/27795334

Index

A
Analytic theory, 62, 206
Analytic thinking, 141–142
Archimedes' law of levers, 20, 24, 51, 64
Aristotle's physics, 4, 76–78, 203
Auxiliary assumptions, 50
Avery-McLeod-McCarty Experiment, 59

B
Balance Theory, 100–102
Behavioral psychology, 89, 90

C
Canons of proof, 13, 16
  method of difference, 16, 20, 24
Centrality, 108
Chiefdom
  coercion-based, 159
  indirect coercive, 163–165
Coercion, 119, 131–133, 144, 193
  Blau, 131
  direct, 131–133, 159, 165, 170, 193
  indirect, 160, 165, 192, 193
  property rights, 131
  rates of coercion, 119, 133
  state monopoly, 173
Coercive power structure, 132
Cognitive model, 100
Collective action, 130
  problem, 34, 107, 108, 144–148
Context complexity, 8, 50
Correlated bias, 26
Crowdsourcing, 49

D
Decision procedures, 84, 152, 157
Deductive explanation, 3, 6, 38, 89, 90
Deductive nomological, 36
  scheme, 208
Design of Experiments, 2, 14, 21, 23, 52, 67
Determinism, 5, 83, 204
Digraph, 101

D. Willer, P. Emanuelson, *The Logic of Theoretical Sociology*,
https://doi.org/10.1007/978-3-031-91709-7

Direct coercion, 193
Dominant strategy equilibrium, 104
Dynamic models, definition, 7

E
Economic man, 6, 95, 96
Economic theory, 6, 31, 95
Einstein's theory, 27, 38, 41–42, 207
Elementary Theory, 7, 55, 64, 107, 137, 147, 150–154, 157, 160, 165, 182, 189, 204, 207
Empiricism, 1, 9, 16, 28, 33, 201
Equiresistance equations, 122–124, 126, 129
Exchange, 64, 85, 129, 132, 152, 172
  social *vs.* economic, 122
Exchange ratio equilibrium, 85, 151
Exclusion, 65, 125–127, 129, 151
  definition, 125
  Marx, 65, 127
Experimental method, 14
Experiment design, 24, 51, 206
Experiment replica, 8, 75, 140, 154
  definition, 53
Experiments, in psychology and economics, 61

F
Falsification
  cycle, 62
  Popper, 38, 62
  as a process, 62
Feudalism, 5, 84, 92, 166, 172, 175, 177–184
File drawer effect, 66
Fisher-design experiments, 14, 20–24, 51–52, 57, 60, 62, 66, 141, 142, 202
Fisher, Ronald, *see* Fisher-design experiments
Fundamental and derived measures, 207

G
Galileo's motion, 4, 20, 24, 51, 77–82, 87, 95, 136, 203, 204
Game theory, 7, 23, 41, 61, 63, 96, 102–106, 108, 142, 157
  applications, 57
Generalization, 2, 15–20, 22, 28–29, 34, 52, 67, 207
  of theory, 101
General knowledge, vi, 2, 9, 22–24, 28, 52, 62, 75, 99, 117, 139, 155, 202, 208
Geodesic, 108
Geometric representation, 6, 77, 78, 80, 87, 95, 203, 205
Gervais and Norenzayan study, 141–142
Graph-theoretic power index, 113
Gravitational waves experiment, 50
Griffith Experiment, 57

H
HARKing, 50
Hempel, C. G., 3
Hershey-Chase Experiment, 59
Heuristic fertility, 36
Historical explanation, 5, 8, 92, 157, 159, 204
Homans
  covering law, 6, 89
  worldview, 204
Hybrid-design experiments, 4, 22, 51, 61
  definition, 57

I
Ideal types, 81, 87, 177
Inclusion, 127–131, 152
  definition, 127
Indirect coercion, 160, 163–165, 192, 193
Induction, vi, 1, 2, 4, 13, 15, 17, 18, 20–23, 38, 52, 61, 62, 117, 139, 158, 202, 203
Inductive method, *see* Induction
Interstate equilibrium, 8, 160, 194

K
Kuhn, Thomas, 41–44

L
Logic of Scientific Discovery, 2, 22, 139

M
Marx, Karl, 65, 93, 127
  power, 55
  worldview, 4, 85, 204
Measurement, scientific, 18, 207
Mechanism, 64, 75, 84, 99, 136, 157, 165, 172
  definition, 55
  example, 53, 55, 59
Merton, Robert, 3, 34–35, 44, 66, 202
  balkanization of sociology, 88, 154
Method of difference, 16, 20, 24
Micro-macro problem, 107
Middle range theory, 3, 34, 66, 154, 202
Mill, J. S., 17
  canons of proof, 13
  definition of induction, 2
  method of difference, 16
Modeling procedure, 64, 75, 99, 140, 149, 154, 165, 205

N
Nash equilibrium, 104
Negotiation processes, 119, 126, 129
Negotiation set, 121, 134, 206
Newton's mechanics, 4, 15–16, 19, 20, 27, 38, 42, 51, 76–82, 87, 136, 155, 203
Normative man, 31
Null coercive structure, 153, 175
Null connection, 124, 125, 128, 131, 152
Null hypothesis, 21, 23, 26, 28

O
Open Science, 50
Open Science Collaboration, 49
Ordering, 127
  definition, 129

P
Paradigms, 41
Patrimonial rule, 171
Pearson, Karl, 20, 28, 37
  product-moment correlation, 13
Philosophy of science, 32, 202
Popper, K., 3
Power condition, 8, 131, 170
  centrality, 65, 125
  exclusion, 65, 125–127, 129, 151
  inclusion, 127–131, 152
  null, 124, 125, 128, 131
  ordering, 125
Power-Dependence Theory, 114
  operant psychology, 111

Primitive accumulation, 5, 84, 93, 94
Prisoner's dilemma, 103, 107, 134
Privilege game, 134
Property
privatization, 94, 165
rights, 131, 176, 178
Prototheory
Balance Theory, 100–102
definition, 99
Game Theory, 102
Rational Choice Theory, 106
Social Network Analysis, 108
Publication bias, 50
Publish or perish, 49

R
Random assignment, 17, 20, 23, 24, 27, 61, 139
Random error, 27
Random selection, 24, 52, 133
Rational Choice Theory, 106–108
Rationality, 41, 55, 57
in economics, 92
parametric, 6, 96, 102, 119, 133, 157
strategic, 6, 55, 96, 102, 119
Reflective actors, 122, 148, 161, 165
Regularity and irregularity, 13, 19, 28, 139
Replication
in experiments, 4
over theoretical scope, 4
Replication crisis, vi, 49
economics, 49
psychology, vi, 49
Research methods
affirming the consequent, 2
survey research, 2, 7, 14, 27, 91, 108, 155
as taught, 2, 31, 201
Resistance equations, 64, 121, 125, 127, 132, 153, 173

S
Sanction, 118, 122
definition, 64
negative, 55, 118
positive, 55, 118
Scientific discovery, 59
Scientific explanation, 208
Scientific revolution, 41
Scientific theory, components, 75
Scope replication, 4, 65–66
Second order expectation, 147
Second order free riding problem, 144
Social action, 95, 96
definition, 119
in Elementary Theory, 117
in Game Theory, 104
Social actors, Elementary Theory, 7, 117–122
Social Darwinism, 31
Social Identity Theory, 142
Social power, 55
Social relation
coercion, 119, 131–133
exchange, 5, 64, 85, 122–126, 129, 132, 152, 172
Social structure, definition, 87
Social theory, determinism, 5, 204
Statistical power, 50
Statistical testing, 14, 23, 26, 139, 141
Status Characteristics Theory, 3, 8, 146, 148, 155, 157, 205
Status-faction model, 149, 163, 188
Status-lineage chiefdom, 161
Status-lineage structure, 166, 190

definition, 149, 161
downward mobility, 167
patrilineal, 161
Status rivalries, 169, 170
Strategic Analysis, 133–135
Strong coercive structure, 154, 173–175
Structural individualism, 106, 110
Structural power, definition, 125
Survey research, 2, 7, 14, 27, 91, 108, 155
Synthetic replica, 158, 206

T
Theoretical model, 8, 53, 61, 140, 150–153
Theoretical science, 33, 201
Theory
construction, 33, 62
development, negation, 203
generalization of, 81
scope, 1, 3, 7, 34–36, 38, 42–45, 51
Theory-designed experiments, 4, 20, 24, 51, 57, 62, 65, 93, 139, 140
Theory-driven experiments, *see* Theory-designed experiments
Elementary Theory, 150, 154
example, 59
external validity, 146
Status Characteristics Theory, 148
Status-Factions Theory, 148–149
Status Theory of Collective Action, 144–146
Theory toolbox, vii, 3, 8, 35, 94, 132, 141, 171–172, 184, 195
definition, 158
Trade negotiation, 192

U
Uncertainty in status relations, 148

V
Variance explained, 49
Vassalage, 177–180, 183
Vulnerability, 112

W
Weber, Max
pluralism, 6, 86, 88, 204
social action, 5, 86, 103, 119, 205
Worldview, 4, 6, 203
Aristotle, 76
Coleman, 6, 93
definition, 76
Galileo, 81
Homans, 6, 89
Marx, 82
Mill, 16
Pearson, 17
Weber, 85

X
Xiongnu, 186, 190–195

The manufacturer's authorised representative in the EU is Springer Nature Customer Service Centre GmbH, Europaplatz 3, 69115 Heidelberg, Germany. If you have any concerns regarding our products, please contact ProductSafety@springernature.com

Printed and bound by CPI Group (UK) Ltd, Croydon, CR0 4YY
01/10/2026
02225852-0001